THE TELEPHONE'S FIRST CENTURY —AND BEYOND

*Essays on the Occasion
of the 100th Anniversary of
Telephone Communication by
Arthur C. Clarke,
Michael L. Dertouzos,
Morris Halle,
Ithiel de Sola Pool,
and Jerome B. Wiesner*

*Preface by John D. deButts
Introduction by Thomas E. Bolger*

*Published in Cooperation with the
American Telephone & Telegraph Company*

THE TELEPHONE'S FIRST CENTURY– AND BEYOND

THOMAS Y. CROWELL COMPANY
Established 1834 / New York

Designed by Joy Chu

Manufactured in the United States of America

Library of Congress Cataloging in Publication Data

Main entry under title:
The Telephone's first century—and beyond.
"On March 9 and 10, 1976, a centennial symposium was
held at the Massachusetts Institute of Technology in
Cambridge, Massachusetts, to celebrate 100 years of
telephone communication . . . This small volume presents
the introductory remarks and overviews of the symposium
papers."
1. Telephone—History—Congresses. 2. Tele-
communication—Social aspects—Congresses. I. Clarke,
Arthur Charles, 1917–   II. American Telephone and
Telegraph Company.

HE8735.T44   384.6   76-30477
ISBN 0-690-01485-6

1   3   5   7   9   10   8   6   4   2

PREFACE
John D. deButts

John D. deButts is Chairman of the Board of the American Telephone & Telegraph Company.

On March 10, 1876, Alexander Graham Bell succeeded for the first time in transmitting the human voice over a pair of wires. From this achievement has developed the wondrously complex telephone system as we now know it in America and around the world. Today, after a hundred years of intensive research, innovation, and development, Bell's invention has made possible the Bell System and over 1000 interconnecting independent telephone companies which provide reliable, swift, and economical communication services that enable several hundred million people to speak with each other and to carry on the great variety of activities essential to the life of a highly developed nation. Indeed, it seems fair to say that life in America as we know it today would not be possible without the telephone.

Despite the fact that the telephone has become a pervasive and integral part of everyday life, the role of the telephone and the telephone system has been recognized by occasional observations, in popular songs, and in the theater, but it has been neglected as a topic for systematic study. On the completion of 100 years of development it is therefore appropriate that we take a look at how far we have come and where we are headed in the second century of telephonic communication.

On March 9 and 10, 1976, we took this look in co-operation with the Massachusetts Institute of Technology. On their campus in Cambridge a highly select group of international authorities was called together to make appraisals of the past, present, and future of telephone communications. Present to hear their views were several hundred persons representing academia and the worlds of business and government.

This two-day summing-up was preceded by a series of workshops involving many scientists and others in three broad areas. During 1975 and the early months of 1976 these workshops heard papers and discussed in depth:

• The Social Impact of the Telephone
• Language and Communication
• Future Impact of Computers and Information Processing

The papers produced in these sessions will eventually be published by the M.I.T. Press.

This small volume presents the introductory remarks and summary statements made by the chairmen of the workshops in the final sessions of the convocation. As such, they are historically important statements and should be preserved as a memorial of this Centennial Convocation. I hope this volume and eventual publication of the workshop papers will stimulate interest in socioeconomic studies of telephone communications. I am, therefore, glad to recommend this volume to all who have an interest in how the telephone and the telephone system has contributed and may in the future contribute to our nation and American life.

CONTENTS

CONTENTS

INTRODUCTION
Thomas E. Bolger

Thomas E. Bolger is Executive Vice President of the American Telephone & Telegraph Company.

A century ago, an event took place in Boston that has since influenced the course of the world in countless ways.

A young man, a teacher of the deaf named Alexander Bell—who preferred to be called by his self-conferred middle name, "Graham"—somewhat unexpectedly accomplished a feat of invention he had been attempting for months—the transmission of intelligible speech on an electric current.

With Bell that evening of March 10, 1876—in his attic workshop—was another man, then even younger than Graham Bell. He was Thomas Watson, a mechanic who built models for inventors and who, in helping Bell, became the first person ever to hear the human voice transmitted by wire.

Fifty years ago, in 1926, Tom Watson recorded his own clear recollection of that momentous evening a half-century earlier. He was *there,* and these are his words: ☞

I first met Alexander Graham Bell in 1874 when he came to the shop in Boston where I was working to have his harmonic telegraph constructed. The work was assigned to me.

During the months that we were working together on his telegraph, Bell often spoke to me of another invention he had in his mind—it was the telephone. I remember my surprise when he first told me that he expected soon to be able to talk by telegraph, explaining to me his conception of an electric current that would copy the vibrations of speech. Guided by the light of his own wonderful theory, Bell found the road leading to the realization of his speaking telephone idea.

On the evening of March 10, 1876, Bell sat in front of the new transmitter in the back room which I had made into a laboratory for him on the top floor of Number 5 Exeter Place, Boston, and I went down the hall to the front room to listen for the results with a telephone receiver. But as Bell was about to speak into the new instrument, a motion of his arm upset over his clothes a battery jar of acidulated water.

In the excitement of the accident Bell called out to me, "Mr. Watson, come here. I want you."

The big mouthpiece picked up his call for help, and I heard every word of it through the receiver at my ear. The new transmitter was better than we had expected or had dared hope.

Bell forgot the accident in his joy over the success of the test that filled the entire evening. He would have brought his great invention to the point where his future was assured. And this first sentence ever sent through the telephone—although apparently very commonplace—was really highly significant. For the telephone today renders no service more vital than in times of emergency, and the first message it ever transmitted was unquestionably an emergency call.

Years passed. On January 25, 1915, the New York–San Francisco line was dedicated to the public service. Dr. Bell and I were invited to take part in this dedication. On this occasion the voice of the chief magistrate of the nation was, for the first time in our history, transmitted from the White House at Washington—3600 miles from the shore of the Pacific. Over this line the words of President Wilson were carried with the speed of light across rivers, plains and mountains to San Francisco; where, in clear tones I received the President's message of appreciation.

That was the transmission of speech for the first time from the Atlantic to the Pacific. I also heard the President converse with Dr. Bell in New York, congratulating him on the consummation of his long labors and distinguished achievements. On that day was accorded to Bell the privilege so often denied to those who advance the world by their discoveries. He had lived to see the full triumph of his great idea.

On that day, though separated by the continent, Bell and I talked to each other. I recognized the familiar tones of his voice and understood without difficulty every word he said. And even when he spoke into a reproduction of his first telephone, his voice brought clearly to me the words of that now historic first sentence, "Mr. Watson, come here. I want you."

Alexander Graham Bell has passed away, but his

> successor shall carry forward to ever greater tri-
> umphs the work which he began, and as long as
> man exists the art which Bell created shall endure.

As Watson noted, it was one of those marvelous co-
incidences of history that the very first intelligible
telephone transmission was a call for help. Certainly it
impressed him, in hindsight, as a symbolic start for the
great advance in communications technology that bene-
fited mankind so much within his own lifetime.

It was also an advance that has served man enormously—and perhaps predominantly—in an economic
and social sense. Not to mention the simple human
pleasures of using the telephone merely to chat with
one another!

The pace of that advance rather quickly outstripped
the technical skills of Watson, the young mechanic, although he personally manufactured the first telephones
and headed, for several years in the infant telephone
business, the Research and Development function much
later vested in the Bell Laboratories.

And Alexander Graham Bell, himself, soon turned
his inventive mind and humanistic concerns away from
the problems of telephone development. But he had
given the world a technological gift of great significance. Even today, 100 years later, its social and economic ramifications are too little understood, and its
potential for the future remains practically unlimited,
only vaguely foreseen in our imaginings.

That is the essential reason why we in the Bell System chose to support both a two-day convocation of diverse and distinguished scholars and the preparatory
studies conducted under the auspices of M.I.T. over
the preceding year. In the studies and analyses leading
toward the two days of presentation and discussion—
as well as in the scope of communications concerns,

both technical and humanistic, spanned by the effort—
we believe a contribution of enduring value to society
can be the outcome.

We believe, in short, that the Bell System and the
academic community seek answers nowadays to much
the same questions. *We* represent an industry somewhat bewildered not by the science and technology of
communications, but by basic, human factors in the
transfer of ideas and meaning. And scholars of the human condition in communications, represent our best
hope of isolating those factors, of finding out exactly
how that transfer of understanding occurs, quite apart
from the technology of transmission.

The international assembly culminating the study of
just such questions is, we think, the most appropriate
of all possible ways to mark the one hundredth birth-
day of the telephone.

But there is another and certainly important aspect
to our corporate support for academic study of communications. And I would be less than frank were I not
to acknowledge it.

That reason is the Bell System's earnest desire for
the widest possible *public* understanding of the communications process. Not public understanding of the
technology involved—which always grows more eso-
teric, more complex—but, rather, a broader understanding or appreciation of how the inherent complexity is organized and managed, of how so basic a human
need has been served so well for so long.

We are now at a time in the development of the
transmission system itself when the *use* of that facility
raises major public issues.

In the contention which currently surrounds questions of vital public interest on the proper modes of
economic organization to provide communications services—and on proper social controls over them—the

Bell System's basic concern is that the public itself not be subject to too much politicking and too little sound, impartial study of the communications process. In all its aspects, it is too important to the public to be distorted by the whims of transient administrators, or in the name of pedantic or dogmatic theories.

Personally, I am convinced that technology today is the least troubling aspect of our concern for the future of communications. It is, rather, misunderstanding of the process at many levels: psychological, social, political and economic. As the science leaps ahead, the art stumbles. We trip on our own ignorance of fundamental relationships which must be mastered to achieve real progress rather than mere expansion of facilities.

The challenge of our time—and it *is* a challenge—is not to develop hardware but to learn hard truths that govern our ability as *humans* to communicate effectively for mutual understanding, if not agreement.

Indeed, I think the best word-picture of one enduring barrier to such understanding was painted in an essay written many years ago by Walter Lippmann. He called the problem "The World Outside and the Pictures in Our Heads."

I think that's a good deal more communicative than "precognitive dissonance," or some other term which in itself limits wide comprehension—and illustrates so distressingly well the problems we face even in communicating about the communications process.

Lippmann's great simplification of the ancient observation that each of us perceives truth—or expresses it —in our own very personal frame of reference, points out clearly this continuing, underlying conundrum in human understanding.

Despite all the marvelous machinery of communications that now exists and will be extended in the future, the process of understanding one another continues to founder on innately *human* barriers. And the technical achievements we pursue are less meaningful to the extent they more effectively transmit only imprecise understanding.

But I do believe that each step in technology which makes the *means* of communication more available, more efficient and more versatile lessens the opportunities for misunderstanding and reduces the barriers to comprehension. Whether those barriers are created by disparities of language, culture, custom, race, nationality—or by any variation of the human experience— ongoing improvement in the technology of communications can't be anything but a positive force in helping to batter them down.

This view of mine does not imply, I hope, any sense that the ideal society—from a communications standpoint—would be an anthill of common purpose and agreement. Dissent and disagreement are not necessarily the product of *mis*understanding. But I think it goes without saying, perhaps, that society's greatest danger at any level of conflict lies not in the tensions of rational discourse. Rather, danger lies in emotional reactions provoked by misunderstanding, prejudice or ignorance—by the simple failure of the symbols of meaning to be interpreted correctly.

This is where the science and the art of communications must be made to mesh more effectively than we fallible "experts" in the process have thus far managed —in the industry, in governments, in the academies of learning.

So, to say it once more, it is what we *don't* know how to do in communications that the Convocation was designed to explore in some depth. The academic disciplines and the purposes of the Bell System can, I hope, combine in a commitment to the future of communications, to the beginning of a dialogue between business and the academic world. Neither is sufficient unto itself, I think, in our times.

But our explicit intent in co-sponsoring the effort is to promote better *public* understanding of the communications process as a whole. And that will be a fitting tribute to the happening of a March evening in Boston a century ago. It was a happening born out of Bell's search to remove barriers to human solitude in his work with the deaf. And *this* happening—in 1976— is in effect a search to break down barriers that still exist in the discourses of mankind.

THE SOCIAL EFFECTS OF THE TELEPHONE

Ithiel de Sola Pool

Ithiel Pool is an international authority on political
communications and behavior, and Arthur and Ruth B.
Sloan Professor of Political Science at M.I.T. After his
education at the University of Chicago he came to
M.I.T. in 1952 as Professor of Political Science and
Director of the International Communications Program
of the Center for International Studies. He worked
closely with the late Max F. Millikan and has been a
pioneer at M.I.T. and elsewhere in the use of quantitative methods and computer technology in political science. He has clearly been in the forefront of the social
and political impact of communications technology, dealing with problems that have interested legislators ☞

9

and other decision-makers. Among his concerns have been privacy and data banks, the effects of television on children, the potential of cable television, and the ways in which technology can facilitate greater citizen communication and participation.

In 1905 someone wrote:

> With a telephone in the house, a buggy in the barn, and a rural mailbox at the gate, the problem of how to keep the boys and girls on the farm is solved.

At the time that he wrote, 34 percent of the American work force were farmers. Little did he realize that 70 years later only 4 percent of the work force would remain on farms. That only goes to prove how easy it is to be wrong.

It is not only forecasters who go wrong. Understanding the past is almost as hard as forecasting the future, for understanding the past also requires a kind of forecasting. To understand the past one must posit counterfactual alternatives. What do I mean by that? One asks the question, What would have happened differently if the course of history had taken a different branch?

So a historical evaluation of the social effect of the telephone requires a forecast about an alternative future. One has to ask what would society have been like if the telephone had not been invented.

As we were starting our research on the telephone's impact, about a year ago, I made such a "what-if" assertion. I wrote that the telephone, along with the automobile, was responsible for that dispersion of population from central cities to suburbia and exurbia. Now, months of research later, and having heard Jean Gottman's paper, I realize that I was just as wrong as the 1905 forecaster of the boys and girls staying on the farm. It turns out that one of the major early effects of the telephone was to make the skyscraper possible. John J. Carty, the chief engineer of AT&T made the point in 1908.

> It may sound ridiculous to say that Bell and his successors were the fathers of . . . the skyscraper. But wait a minute. . . . Suppose there were no telephone and every message had to be carried by a personal messenger. How much room do you think the necessary elevators would leave for offices. Such structures would be an economic impossibility.[1]

In papers prepared for the symposium by Jean Gottman, Ronald Abler, J. Allen Moyer and Suzanne Keller, the complicated relationship between the telephone and urban growth in twentieth-century America has been explored. Before the coming of the telephone, business neighborhoods were walking areas. All the traders of a particular kind would congregate in a few blocks, in order to be able to walk up and down the street to do business with each other. Rents in these dense centers of activity were very high. The telephone, the streetcar, and the skyscraper provided an option; they made it possible for firms to escape to cheaper and more commodious quarters. After the 1880's one could escape either by moving *out* or moving *up*.

In other ways, too, the telephone contributed to the growth of downtown business districts. In the early years, before telephone lines were readily available everywhere, the phone companies supported the zoning of American cities so as to stabilize planning for service areas.

Thus we find that the effect of the telephone on urban patterns might have been the opposite when there

---

[1] In John Kimberly Mumford, "This Land of Opportunity," *Harpers Weekly*, Vol. 52, August 1, 1908, p. 23.

were few phones from what it became when there were many. Today when Vail's goal of universal service has been achieved, there is much speculation that, for the first time in history, social organization may cease to follow the lines of contiguity. Suzanne Keller has pointed out that communities of people who share common interests begin to emerge without regard to location. Ronald Abler has presented historical evidence of the decreasing importance of contiguity in the organization of American business. Bertil Thorngren, from Sweden, however, has cited empirical evidence that relationships between people tend to survive when both telephone contacts and face-to-face contacts occur with the same individuals. And so, he argues, as do Gottman and Mayer, that contiguity will continue to be a powerful force in the formation of human communities.

Before going further, I should, perhaps, explain how our symposium on the social effects of the telephone proceeded. In a series of five sessions held since October 1975, about 20 papers were presented covering various aspects of the social impact of the telephone.

The largest group of papers, eight in total, dealt with alternative paths of development which in the early years might have been anticipated for the telephone system. One of the alternatives that came to mind to many people a hundred years ago was to create a system on telephone wires similar to what we today know of as broadcasting. The papers by Sidney Aronson and Asa Briggs both presented examples of what Briggs called "The Pleasure Telephone." In August 1876 the British magazine *Nature* described Bell's invention and its future uses:

When we are going to have a dancing party, there will be no need to provide a musician. By paying a subscription . . . we can have . . . a waltz, a quadrille, or a galop, just as we desire. Simply turn a bell handle, as we do the cock of a water or gas pipe, and we shall be supplied with what we want.

Starting in 1893, in Budapest, for some decades there was a system with news, and music and ads in a pattern essentially like that of present-day radio.

There were good technical and economic reasons, recognized by Bell and his associates, for choosing to develop telephony as a point-to-point network rather than, as in Budapest, as a one-way broadcasting system. The cost and poor quality of sound in those early years made entertainment not a profitable prospect. On the other hand, the switchboard, as Colin Cherry pointed out, made a point-to-point network economic, for uses for which people were ready to pay.

Bell and Hubbard decided to provide a public service for hire, enabling people to talk to each other at will. They chose to do that rather than to sell telephones as retail appliances or to broadcast programs to mass audiences. That choice had profound social consequences. It made the telephone system into an important democratizing force—what Cherry has called a creator of mobility and change. In contrast to mass media, which mobilize a population, the telephone network was a technology suitable to America's pluralist populist society. While a telephone network may strengthen the cohesion of a society, it also breaks taboos of social access. The person answering the ring of the telephone does not know the status of the person to whom he is about to talk.

Both Cherry and Jacques Attali suggest that where the telephone was offered as a means of disseminating access, it spread rapidly because it was well adapted to meeting that need. Where, as in France, telecommunications were seen as a facility for the regime to disseminate its own messages, the telephone spread slowly and its growth was inhibited.

One set of papers in our symposium dealt with the role of the telephone in modern life. The telephone entered American life softly, in a way incongruent with the loud and imperious ring with which we associate it. Sociologists and historians hardly took note of how people used it. Writers did a little better. John Brook uses the mirror of literature to describe the changing use of the phone as it went from a magical novelty to a pervasive mode of human interaction.

We have all lived through being "all alone by the telephone waiting for a call" and have experienced the comfort of being able to talk to a friend or relative far away. It is often easier to talk on the telephone protected by separation when relations are strained. The endless calls by adolescents are part of the stereotype of American family life.

One way to recognize what the telephone means in these ordinary passages of daily life is to look at what happens to people in those rare situations when they are deprived of their telephone. Last year New Yorkers were interviewed who were deprived of service by the exchange fire. People without access to the phone felt insecure; the phone was a link to "help" if needed. It allows one to know, for example, that the baby with the sitter at home is well.

Brenda Maddox has noted the important part played by the telephone company in the changing role of women. For millions it has helped overcome loneliness and isolation. For hundreds of thousands in the days of the manual switchboard it provided interesting, respectable work. In their availability, operator jobs, along with teaching, nursing, and secretarial work, were at an earlier stage major opportunities for women to come into the world of work. Today we no longer accept the notion of such protected enclaves, but in their day they served a purpose.

Another group of papers looked at the telephone conversation as a special form of human interaction. Alex Reid compared the kinds of tasks that are done well on the telephone with those that require face-to-face interaction. Imparting of most kinds of information is done at least as well if not better without the presence of sight. On the other hand, persuasion may be harder to achieve and the gratification of the participants may be less in the absence of personal contact. Emanuel Schegloff has examined in detail the games people play as they start a telephone conversation. For years people were told, for example, in etiquette books, to start a telephone conversation by identifying themselves. People rarely do it. "Hello." "Hello." "How are you?" "Fine, Jim, how are you?" is the way people play the game, giving as little information as they can till each is assured that he is recognized by the other.

Other papers examined special uses of the telephone. Paladagu Rao described telephone education, for example, for shut-ins. David Lester described the experience of hot-lines for suicide prevention, for drug addicts, or for rape victims.

The papers in the symposium only lifted the curtain on the vast social effects that have been produced by Graham Bell's invention. It is extraordinary how small is the literature on the social impact of the telephone. If you look in the library at the books on the railroad and its impact on the development of the modern world, at the literature on broadcasting and its impact on youth, you will find title after title, shelf after shelf. Even if you look at such an obscure subject as the longitude problem in the seventeenth century and its social consequences, you will find much more literature than there is on the telephone.

Perhaps there are some good reasons why. The telephone has been an unusually benign invention. One almost never hears of a serious accident or fatality caused by the telephone. Furthermore, people choose how they use the phone; it is not outside their control in the way that one-way media or traffic are. So, not feeling exploited by it, people take the telephone for granted.

There is another reason, too, why it is not easy to identify the social effects of the telephone. Unlike many other inventions, the telephone has few constraining effects on society. It is a facilitator, allowing people to do more easily whatever it is that they decide to do. Consider railroads, in contrast. When a railroad was laid, its physical location had a constraining influence on where to locate factories, on which cities grew, and even on the peace and quiet of neighborhoods. Or consider television. We as individuals have almost no influence on the programs offered—though in the mass we do determine them through the ratings. But as individuals we care about our individual preferences, not about the average of which they are a part. So we feel constrained and are influenced by what is offered.

The telephone has been a remarkably adaptive device. It can be used for whatever we choose to use it. In that sense it has been a contribution to freedom. Its effects, therefore, are as many and as varied as the uses to which people put it.

One can sense that fact from the many respects in which one finds opposite or contradictory effects of the telephone. On the one hand it has contributed to the growth of skyscrapers, on the other hand to the growth of suburbia and exurbia. On the one hand it has helped police to catch criminals; on the other hand it has helped criminals evade the police. On the one hand its ringing invades our privacy; on the other hand in the telephone era it is no longer acceptable to drop in on people without warning them by phone in advance. It has really added greatly to privacy.

The telephone, in short, has expanded human freedom. What can be harder than defining the consequences of freedom? A deterministic model of cause and effect is bound to misperceive them.

The most difficult problems in the social sciences concern the understanding of free behavior, or in other words, the understanding of choice. Choices are made, for example, in elections or markets. People use these in many ways to achieve many different outcomes. Yet there are important things to be said about them as influences on the outcome, even though the outcomes are not single-valued. So, too, there is much to be said about the social effects of the telephone, even though the first thing one has to say is that the existence of a universal, geographically unbounded, instantaneous means for people to communicate with one another, opens up an enormous range of alternative paths of human development.

The citizen may be justified in taking the telephone for granted in his daily life. It is something that is going to be there to serve him. For the historian or social scientist, however, it would be well to pay more attention to that remarkable device that has so profoundly shaped our lives.

Interestingly enough, to date, the best social analysis of the telephone and what it could do was by the very people who invented it and developed the system. Alexander Graham Bell, Gardiner Greene Hubbard, his father-in-law and financial backer, and Theodore Newton Vail showed more insight into what the telephone system could mean and how it could best serve society than anyone else I can think of. It was not foreordained that the telephone was developed as a public service rather than as a business machine sold to the few. It was not inevitable that the goal from the beginning should be taken to be universal service in every home and office, and a linked network reaching throughout the world. Those paths of development were the results of deliberate choice. The telephone was born out of a social goal—to understand and cure deafness. It was developed with clear-sighted analysis of social needs and how the industry could be organized ultimately to meet them.

The men who built the phone system were all social idealists. Bell was a teacher of the deaf. Vail had come from public service and understood that the industry that he was building was a public service. Hubbard had earlier been a critic of the telegraph industry for its failure to meet social needs. They were quite clear about their goals of integration of society and universal service.

The clarity of Bell's perception and forecast is best conveyed in an oft-quoted letter that he wrote to a group of British investors in 1878, only two years after the invention. I take the liberty of quoting it once more.

The simple and inexpensive nature of the Telephone . . . renders it possible to connect every man's house, office or manufactuary with a Central Station so as to give him the benefit of direct telephonic communication with his neighbors at a cost not greater than that incurred for gas or water.

At the present time we have a perfect net-work of gas-pipes and water-pipes throughout our large cities. We have main pipes laid under the streets communicating by side pipes with the various dwellings enabling the inmates to draw their supplies of gas and water from a common source.

In a similar manner it is conceivable that cables of telephonic wires could be laid underground or suspended overhead communicating by branch wires with private dwellings, Counting Houses, shops, Manufactuaries, etc., etc., uniting them through the main cable with a Central Office where the wires could be connected together as desired establishing direct communication between any two places in the city. Such a plan as this though impracticable at the present moment will, I firmly believe, be the outcome of the introduction of the telephone to the public. Not only so, but I believe that in the future wires will unite the head offices of Telephone Companies in different cities and a man in one part of the country may communicate by word of mouth with another at a distant place.

I am aware that such ideas may appear to you Eutopian. . . .

Believing, however, as I do that such a scheme will be the ultimate result of the introduction of the telephone to the public, I would impress upon you all the advisability of keeping this end in view that all present arrangements of the telephone may eventually be utilized in this grand system.

This description was offered at a time when there were in the world only a handful of phones and no such thing as long-distance service.

TECHNOLOGY AND THE TELEPHONE

Jerome B. Wiesner

After graduating from the University of Michigan in 1937 Dr. Wiesner began his career in Broadcasting Service at the university. In 1940 he became Chief Engineer for the Acoustical Record Library of the Library of Congress. During World War II he worked in the Radiation Laboratory at the Massachusetts Institute of Technology and was a distinguished member of the faculty from 1946 to 1971. During this period he held a wide variety of administrative posts, including an appointment as Special Assistant to the President of the United States on Science and Technology, 1961–64. He has been President of M.I.T. since 1971, has written on scientific subjects, and is on the governing boards of numerous scientific and civic organizations.

During the U.S. Bicentennial year, M.I.T. carried out workshop studies on several of the significant technical-social aspects of the world the United States enters in its third century. Communications was early identified as one of the aspects to be studied, so we welcomed the opportunity to join with the Bell System to celebrate the Centennial of the telephone, building on both our strengths to organize a convocation on communications in which the telephone Centennial would be celebrated by the scientific and scholarly community as well as by the Bell System.

We hope that our readers enjoy this series of colloquia as much as the participants did in preparing them. It was a privilege to be the keynote speaker at the centennial celebration. My especial feeling of honor went beyond the bureaucratic reasons, to the personal, because I still think of myself as essentially a communications engineer, and I see most of life's situations, particularly those that most confound us, as communications problems.

The discussions during the two-day convocation provided insight into the outer reaches of research that will help us understand and, hopefully, relieve many of these problems. Furthermore, it is generally believed that emerging technologies will greatly expand available communication capacity for voice, video and data, and at the same time reduce their costs sufficiently to open up wholly new modes of interaction, work and means of conducting our personal lives. I welcomed this refreshed course, for I have a Walter Mitty–like dream of myself once again at work on communications problems.

When I was an active researcher, my most rewarding and continuous outside links, starting with World War II collaboration on radar and continuing afterward on scatter propagation, speech analysis, information and coding theory and noise work were with people on the staff of Bell Telephone Laboratories. Historically, M.I.T. and the Bell Telephone Labs have been close, collaborating and competing intellectually though perhaps on a horse- and rabbit-basis, considering the size of Bell Laboratories.

This spirit of collaboration was born with the telephone, in fact. Alexander Graham Bell was a neighbor of M.I.T., known then as Boston Tech—Technology, as he called it. He used the physics laboratory for some of his measurements and used M.I.T. students as helpers. One of the first demonstrations of Bell's telephone took place only a few blocks from M.I.T., in buildings that stood on the site of the laboratory where today Edwin Land studies color vision and spins out his modern miracles. I doubt that he inhabits the same buildings, but you cannot be sure.

The report of the President of M.I.T., John D. Runkle, for 1877 contained a section entitled "Improvements in Telephony," which reported that:

Prof. A. Graham Bell gave a brief sketch of the history of telephony and the various ways in which he had simplified the apparatus employed. He described the original instruments, in which a battery current was used, in one of three varieties, which he called a year ago intermittent, pulsatory and undulatory currents, characterized respectively by the alternate presence and absence of electricity upon the circuit, sudden changes in the intensity of continuous current and gradual changes of intensity of a continuous current, like those in the density of air, occasioned by simple vibrations. His receiving instruments then resembled a human ear, in connection with electro magnets and coils, each terminal apparatus being different from the receiving. He had ascertained that it was simply necessary to have a permanent magnet, with a simple coil with a magnetized iron core, without a battery. The larger the plate, the better the sound seemed to be; he had tried every size and thickness, from that of the thumbnail to a sheet of iron four feet square and from the thinness of tissue paper to a boiler plate a quarter of an inch thick, with nearly the same results: a goldbeater's skin and similar membranes in imitation of the drum of the ear were entirely unnecessary.

The M.I.T. President's report continued on Bell's telephony by observing that:

The vibration is certainly not that of a membrane, and may be called, for want of a better name, molecular. The essentials for the telephonic apparatus as at present constructed are a permanent magnet, a coil, and a plate in front so placed as to vibrate freely without touching, and all these parts he has varied in many ways to size, shape, thinness, material, etc. He finds that the thinner the coil, the better the results and he has had the best effects from a plate of soft iron, four inches in diameter and one thirty-second of an inch thick.

He exhibited a new and portable telephone, having the magnet and coil in the handle; this was passed round among the audience, who were thus enabled to hear music and conversation in a far distant room in the building. He had made a serviceable instrument which he could carry in his vest pocket.

He spoke of the social and business uses to which the telephone might be put and drew an amusing picture of its future applications, as a time- and labor-saving apparatus at a moderate cost and for domestic uses.

He [Bell] concluded his experiments by passing a voice through the bodies of twenty persons, standing side by side on the platform and taking hold of each other's hands. This was overcoming an immense resistance, much greater than that offered by the Atlantic cable, suggesting that, if this resistance be the chief thing to overcome, telephony may yet be substituted for telegraphy across the oceans, as it soon will be across the continents.

Bell was the storybook inventor of old—very poor and very driven. He went for considerable periods of time with no income, with little support, neglecting his speech school in order to have the time to work on his invention. The records don't indicate whether or not our academic predecessors occasionally treated him to a warm meal. I hope that they did. Incidentally, an M.I.T. catalog for that period listed problems in telephony under "studies in electricity."

As I prepared for the M.I.T. two-day convocation, I wondered how Alexander Bell would have reacted if he could have been with us. Would he have understood the scientific discussions? Would any of our talk have seemed relevant to his goals and interests? Yes, I believe that they would have and that he would have found the celebration of his invention both interesting from a scientific point of view and rewarding, even a source of pride in spite of his modesty, because the industry and company he founded have thrived and have been the source of immense social and economic good.

The papers prepared to highlight the impact of 100 years of the telephone and of research related to communications are a special distillate from the extremely broad range of all such activities. The three groupings, (the) social impact of the telephone, (the) new approaches to a realistic model of language, and (the) future impact of computers and information processing, represent extremely active fields and are particularly suited for a future-oriented Bicentennial.

Bell would find the scientific sessions particularly interesting, for many of them are lineal descendants of his earliest experiments in speech and hearing. Others reflect the interests of his later years. He would be interested in modern linguistics theory linking language and cerebral structures, and modern instrumentation for studying hearing and for synthesizing speech, for these would take him back to his youthful experiments. He would be interested in the computer programs for decoding spoken language, and no doubt would regard the modern visual speech presentation devices as great improvements over the smoked cylinder which he used for studying sound. He would no doubt go away feeling the need to revise his little paper on visible speech. He would be a bit overwhelmed by the power and universality of computers, but not for long. He would perhaps be most excited, as some of us are, by the coming together of the theoretical ideas of linguistics, the neuro-sciences, psychology and communications science, hopeful that they hold the keys to the mysteries of some deep questions of human perception and cognition, which were matters that Bell frequently speculated on in his efforts to understand better how to teach the deaf.

In this world of communications miracles, we perhaps don't appreciate fully the unique position the telephone now holds and, in Bell's view, was destined to hold from the start. Consider its very special properties. It operates in real time, it provides two-way communications and it is, in principle, designed to interconnect all subscribers. It couples distant minds by a strand of wire. Bell's efforts to convert his vague vision of transmitting speech over wire into reality drove him to become a truly interdisciplinary man. From a medical school friend he obtained an ear, carved from a cadaver, which he studied intensively for insights into the process by which it converted acoustical signals into the perception of sound. For a considerable period of time, this was his most valuable research tool and he took it back and forth on his trips to Canada. The diaphragm in the telephone instrument evolved as a result of observations he made on that ear. Bell also had to learn something, just the least little bit, about signals on wires, about materials and, in order to create an effective system, about switching. Progress in communications has been paced by research along these same diverse fronts ever since.

For convenience these efforts could be neatly divided into the study of physical components and systems and the study of signals and information. The object of it all being to handle that somewhat illusive commodity called information; to understand its generation, its perception and the effects on *it* of various forms of *infidelity*, usually electrical, in handling it, including band-limiting, distortion and the introduction of noise. The physical systems proved easier to understand than the commodity being processed, but both were the subject of extensive theoretical studies. Claude Shannon's monumental work on information theory capped the long efforts to understand the nature of the communication process by providing a quantitative definition of information and a relationship between bandwidth, signal-to-noise ratio and signaling rate. During more or less the same period, there developed the theory of feedback and its application to communications control, the evolution of large switching networks and the increasingly powerful computers—including Vannevar Bush's analog machines at M.I.T., the relay machines developed at Bell Laboratories, and those based upon the concepts of the programmable computing machines developed by John Von Neumann at Princeton.

These developments paralleled a period of explosive growth in the neural and cognitive sciences. The concepts and understandings drawn from the communications and computational studies have had a significant role in these developments.

After the successful demonstration of Bell's talking wire, as he sometimes called it, and the issuance of a patent, he and his supporters turned to the creation of a company to exploit the invention, leading inevitably to the development of a telephone system. They sold stock—with surprising difficulty at first, people regarding it somewhat as a toy in the beginning—and then they began the manufacturing, the installation of instruments and the licensing of territories where others could build telephone systems if they used the Bell company's equipment. Then, as now, founding a new business venture was risky and more difficult than it looked. The inexperienced founders of the Bell Company found it difficult to cope with the vast variety of problems they encountered, including patent challenges and patent infringements, the need to improve the technical performance of the early telephone instruments, the need for switching capability and even the default on bills due. In desperation they turned to a young man named Theodore Newton Vail, who proved to be an extraordinary management genius and who guided the rapidly growing company steadily, successfully. Alas, he also moved it from Boston to New York. Incidentally, the Vail Library, which he persuaded AT&T to purchase and present to M.I.T., one of the world's outstanding collections of the history of the telephone and other early communication devices, is a major unit of the M.I.T. libraries rare book collection.

It is commonplace to talk about America's love affair with the automobile. The telephone has not been described—as far as I know—in such romantic terms. Yet, the general dependence upon it is, if anything, greater. The affairs of the nation would certainly stop if the communications system ceased to operate for even a few hours. The telephone provides a sense of support and protection while the automobile affords mobility and escape, so the attachment is of a different nature. Perhaps that is why the telephone system long ago earned the affectionate nickname of "Ma Bell." At the risk of being called a male chauvinist, I suggest that "Pa Bell" should get credit for the system, at least today. The telephone certainly was a major factor in determining the shape of modern society, making possible the diverse living and business styles that have evolved. Could we have had suburban life without the telephone, for example, or centralized business control?

A biography of the telephone, its many technological offspring and its industrial ramifications, would provide an excellent vehicle to carry the story of the changing role of technology in the industrialized nations and of the changing attitudes toward industry and technology. A history of the efforts to improve and extend the telephone is equally a history of the development of industrial research in the United States. It is as well a vivid reminder, much needed today, of the vital role of basic research in human development. It demonstrates how knowledge, accumulated for one purpose, can enrich countless, apparently unrelated, fields, and how a determined and continuing research and development effort spawns new devices, industries and societal benefits.

One might argue that the telephone is a special case and that the social and technological impact was perhaps unique. But the impact of the telephone is unique only in its coherence. I believe that we could demonstrate similar though less easily followed and less dramatic tracks left by other discoveries and inventions, such as the steam engine, organic chemical synthesis, the electric magnet, the typewriter, the loom or the internal combustion engine.

This centennial of the telephone occurs coincidentally with the Bicentennial celebration of our nation's founding, a Bicentennial which finds the nation in a somber, bewildered, questioning mood at a time characterized by a loss of confidence in the traditional institutions of government, industry, and education, and a sense of despair over the future of the country. Much of the concern today is focused on those aspects of the society which in the past accounted for its unique economic achievements and dynamism and, particularly, the impact of technology and technologically based industry on the quality of life.

It is interesting to contrast the mood today with that of the centennial year 1876. That was a time of great excitement and expectation for the individual, for the country, for the development of industry, for the role of science and technology in the creation of an ever more satisfactory future. The centerpiece of the centennial celebration was a great exhibition in Philadelphia featuring industry and technology. It was opened and closed by the President of the United States. Its central piece was a great Corliss steam engine that the President started to inaugurate the exhibition and shut down at the exhibition's close. Not much noticed in the general jubilation was a talking Telegraph exhibited by a Mr. Bell from Boston. This leads one naturally to wonder whether in our present questioning mood we are not overlooking some nascent development which has equal potential for social contribution over the next century.

We who share a responsibility for the continuing development of science and technology, must understand the issues and problems that relate to science and technology—which all do in some way or other—in order to help revitalize the national spirit.

I have reflected on these problems quite a lot in the decade or so that I have become aware of how widespread and growing was the belief that the large-scale exploitation of science and technology was creating more serious problems than it was helping to solve, or at least that the emerging problems were more complex and unmanageable than those they displaced. I have concluded that while there is a valid basis for the concerns, the trade-offs are very much on the positive side. Technology does create new and often more complex problems, but the growing complexity is offset by the breadth, sophistication and magnitude of the capabilities for contending with them. At the same time, people are provided with the benefits of the innovations. I doubt that any one of us would prefer the conditions of 1876 or even those of a typical *traditional*—that is, undeveloped—society today, to those of contemporary life if we were forced to choose.

I am convinced that we must accept this bittersweet aspect of a technological society and expend our efforts learning how to cope with it more effectively. I say this because I believe that a technologically based society is —must be—a dynamic system—a learning system— in a continuing state of change and evolution, requiring new technologies and new organizational forms, new relationships and probably even new life-styles as it evolves.

Many people will find this premise troublesome, for many have been hoping that the world might one day— sooner rather than later—approach a steady state in which change, and especially technologically induced change, would cease. Sad to say, they, and we, must accept the fact that there is not likely to be a stable state in the sense that new problems and new opportunities will cease to arise or new and better solutions found to old problems. I am convinced that industrial societies can only exist in a state of dynamic equilibrium that involves continuing adaptation to the changing too man-made world and to a natural world that is changing as a result of people's actions.

There is no way to avoid being at the same time creatures of nature, completely dependent upon the planet and the sun for our life-sustaining needs, and participants in an on-going social process. We cannot turn off the synthetic world and its demands, or even run the technological clock backward to a simpler time. That being the case, we can only strive to live in harmony with both worlds, realizing that inevitably this will involve continuing adaptation, continuing change, but not necessarily traumatic change. Advances in science and technology will make it possible to do many things better, with less impact on the environment, with less burden and danger to the individual, less expensively, more reliably, so we should expect that most industries and fields will continue to evolve, some, such as the life sciences and the communications and the computer sciences, perhaps explosively.

To a very considerable extent, the ease, orderliness and timeliness with which new technologies will be developed and be adopted is primarily a problem in education, social management and communications.

I would like to make another observation about the technology change. At the very time that there is so much concern about the impact of new technology and the disorienting effects of change—I detect that the character of change is itself changing and, I believe, slowing down. Science and technology continue to advance rapidly, but the societal impact is less dramatic. It is generally agreed that the highly industrialized nations of the world have entered a new phase in their evolution, sometimes described as the "post industrial society." I don't particularly care for this characterization of the situation, for it seems to imply that the technological-industrial phase of social evolution is behind us, that industry as it is now is capable of supplying unlimited quantities of goods to meet all our public and private needs and that the major social problem is to use our present technologies more effectively, propositions that I seriously question. On the contrary, I believe that to cope with the current and future needs for resources as well as the environmental, social development, educational and health needs will require new knowledge, new industrial initiatives and no doubt new industrial institutions.

Nonetheless, something is very different in a technologically mature society. During the first phase of an industrial revolution the new technologies cause major and largely unexpected changes in the situation of the typical citizen of the industrializng nation. Now new technologies are more likely to be put to work to maintain or slightly improve previous achievements, as in the case of the search for new energy sources or energy conservation techniques or the efforts to improve air quality or the development of better data-processing systems, or of much high-capacity communications systems. Now the likelihood of producing violent, traumatic discontinuities by the introduction of new technologies is considerably smaller than in the past, in part because of the maturity of technology, and in part because societies are learning to be on guard against such occurrences. In fact, discontinuities are much more likely to be of a degenerative character, and result from the failure to have new technologies available when needed. The failure, for example, to develop alternative energy sources, or adequate water resources, or safe pesticides, or the failure to develop adequate new materials or raw material sources as existing ones are used up, could have very damaging effects on the economy and conditions of life. Such technological failures could lead to major political discontinuities as well. In my view all the current and foreseeable technical problems that concern people, such as those I just listed, can be dealt with. But for many of them we need better decision-making and social management techniques than now exist. Some political observers are pessimistic about the ability of a democratic state to manage the complex problems of a resource-short industrial society. If you accept their conclusions, freedom itself rests upon our ability to create those adequate new technologies.

Many factors conspire to slow down and even stall what might be called replacement technologies. Perhaps most important among these factors is the high cost and the long time required to bring an alternate technology to the refined state of the one it is expected to replace and which attained its level through many generations of design and redesign. A current example of this is the difficulty experienced in bringing nuclear power to a socially acceptable level of reliability. In that situation we don't even know how to define or measure an acceptable level of reliability. Even when the alternate technology exists and is socially acceptable, the cost of replacing existing plant capacity may be a limiting factor. For example, it would cost many hundreds of billions of dollars to create enough synthetic fuel capacity to replace the present natural petroleum sources. Similar limitations may restrict the rate at which future communications services, including cable distribution of video, can expand.

Also, regulatory constraints or even the possibility of regulatory actions can be major inhibitions to the development of new technology in some areas—for example, in a growing number of health areas, in the nuclear field or in the use of coal for the generation of electricity.

So, I would suggest that the problem for the next half-century is at least as likely to be how to assure change as how to live with it. In any event, I doubt that we are likely to experience future dislocations as dramatic and disorienting as those caused by the shift from farm to city or the social and cultural changes brought about by the development of mass-production industry, by communications technology including the telephone, cinema, phonograph, radio and television, and by the recent progress in medicine and health care including fertility control. Further, the changes in international relations brought about by technology where nuclear weapons, communications and trade have created an interdependence that must ultimately —and the sooner the better—draw the people in the world into a truly cooperative mode.

The key question, then, is not how to direct technological change but how to direct change so that the resulting situation is more satisfying. And as we celebrate the centennial of the telephone, the question is, what can communications technology and the information sciences contribute in that direction?

LANGUAGE AND COMMUNICATION
Morris Halle

Morris Halle was born in Latvia, studied engineering at City College in New York and the University of Chicago, then went to Columbia to study with the great linguist Roman Jakobson. In 1949, when Jakobson left Columbia for Harvard, Halle went with him. Halle obtained his Ph.D. from Harvard in Slavic Linguistics and then became Assistant Professor in the Department of Modern Languages at M.I.T. and a member of the staff of the Research Laboratory of Electronics in 1951. He, along with his colleagues, was instrumental in developing M.I.T.'s graduate program in linguistics and will be the first chairman of the new combined department of linguistics and philosophy in 1977. He has won many honors, including election to the presidency of the Linguistics Society. Among his many books are Preliminaries to Speech Analysis and The Sound Pattern of Russian.

Questions of language are a most appropriate topic for discussion on the occasion marking the centennial of the telephone, for not only is the transmission of spoken language the primary function of the telephone, but also as I shall have occasion to remark later, the telephone's inventor, Alexander Graham Bell, had a deep scientific interest in problems of language and must be credited with an important conceptual advance in our understanding of the sound structure of language.

Since early in 1975, M.I.T. has been host to a series of workshops on problems of language and cognition, which were organized with the help of a grant from the American Telephone and Telegraph Company as one of several such activities leading up to a two-day convocation. The workshops brought together between 20 and 30 workers from different research groups with significant involvements in problems of language. Our thinking, like that of most students of language in this generation, had been influenced by the work of our M.I.T. colleague, Noam Chomsky. Of particular significance to our enterprise was Chomsky's insistence that the proper and overriding aim of linguistic description must be to provide an account of the knowledge that native speakers have of their language. The importance of this focusing on the cognitive aspect of language may not be immediately apparent to the nonspecialist. As a participant-observer of the history of the field during the last quarter century, however, I have no doubt that few things have had such a far-ranging effect on the development of the entire field as Chomsky's insistence that language is a form of knowledge peculiarly accessible to humans and that it is, therefore, akin to other manifestations of our cognitive faculties such as our ability to perform computations, to play games and to invent, plan and execute complicated structures and maneuvers.

If language is knowledge that is peculiarly accessible to humans, then all language must share certain essential features which are especially well matched to the intellectual capabilities of the young child, for in all known linguistic communities command of language is acquired at a very early age. That languages share many substantive properties is by now a well-established fact: every known language forms sentences by concatenations of words; in every language ever studied, words are made of sequences of a restricted number of speech sounds, and these in turn, as Bell was one of the first to see clearly, are complexes of a small number of phonetic properties, and so forth. In addition to such substantive universals, languages also share properties of a more abstract kind, which in the linguistic literature are now designated by the term *formal universals*.

It was one of Chomsky's suggestions that an important formal universal of language is the syntactic transformation, a special computational device first described by Zellig Harris. A large portion of Chomsky's work as well as that of others during the last twenty years has been concerned with establishing the character and the proper role of the syntactic transformation. Initially transformations were assigned a very large role in the functioning of the language. More recently it has become evident that transformations were unsuited to some of the tasks that were assigned to them. This discovery, as Joan Bresnan has noted, elicited different responses from different researchers. Some proposed to overcome the difficulties by increasing the power of transformations; whereas others have followed the lead of Chomsky and opted for a significant limitation on the power and role of transformations. Though participants in the workshop tended to agree with the latter view, there was no consensus about the precise character of the limitations that should be imposed on transformations. A proposal that seems to hold considerable promise has been presented by Bresnan.

Among the tasks that transformations were especially unsuited for was the characterization of relationships among words. Since relationships among words obviously play a big role in a person's knowledge of language, other means will have to be found to express them. The obvious candidate for expressing this information is the lexicon, for it is there that the speaker's knowledge of the words of his language is represented. The information about the relations that a given word bears to any of the others will then be part of that word's entry in the lexicon. The complexity of this information requires a fundamental revision of our conception of the character of the lexicon. Rather than being a simple listing of more or less odd facts, the lexicon must now be regarded as an active device possessing a structure that must be carefully investigated.

By a most fortunate coincidence, renewed interest in the lexicon developed at about the same time also among psychologists. George Miller and Philip Johnson-Laird have just completed their monumental *Language and Perception,* a large portion of which is devoted to an inquiry into the form and function of lexical entries. Miller's description of this work generated considerable excitement as it dealt with topics that had been uppermost in the minds of several of us. It also led to a very fruitful exchange of information about much unpublished work that had been going on here as well as elsewhere. One side product of this was that Ray Jackendoff started to come to our meetings and to take an active part in our discussions. The main results of those discussions are contained in the papers prepared by Miller and Jackendoff.

The issue that provoked the most heated debate among us was the manner in which knowledge of language is utilized in the production and understanding of utterances. While few would disagree with Chomsky's remark that "a reasonable model of language use will incorporate, as a basic component . . . the speaker-hearer's knowledge of the language," there was wide divergence about the precise way in which a processing model of language incorporates this knowledge. The problem might perhaps be clarified by an example, which I have adapted from George Miller's paper. Miller points out that the words *woman* and *person* both designate individuals that are animate and human; they differ in that the word *woman,* in addition, indicates that the individual is female, whereas the word *person* provides no information about the individual's sex. If, as appears plausible, the semantic information just presented has to be used by the speaker in the understanding of sentences, then it might be supposed that the word *woman* is semantically more complex than the word *person,* and this difference in complexity would be reflected by differences in the time it takes to understand such otherwise identical sentences as *John's wife is the woman on the right* vs. *John's wife is the person on the right.* It appears, however, that no such differences have ever been found.

What is one to conclude from this negative result? Some would conclude that this result suggests that the postulated difference in semantic content of the words *woman* and *person* lacks psychological reality. Others have argued that since the difference between the two words can be established by other behavioral tests, the lack of difference in reaction time does not speak to the issue of psychological reality. While with respect to the example in question the evidence appears strongly to favor the latter view over the former, the evidence is less clear with respect to other aspects of linguistic knowledge. There are particular disagreements about the psychological reality of transformations. On the one hand, Eric Wanner and Ronald Kaplan propose that the transformational model should be replaced by a nontransformational model of their own devising. On the other hand, Merrill Garrett and Kenneth Forster argue that the evidence favors a processing model that incorporates a transformational model of the more familiar kind. It is obvious that considerable further work and thinking will be required before a consensus can emerge.

Last but by no means least, the workshops considered also the insights into the nature of language that might be hoped for from observations of language acquisition by young children and of language loss in the brain-injured. Details of these discussions were presented by Susan Carey, by Michael Maratsos, and by Sheila Blumstein, Mary-Louise Kean, and Edgar Zurif. Because of the extreme variety of the data that have been amassed here, great difficulties are encountered in attempting to interpret them properly, and agreement concerning the significance of a given observation is often impossible to achieve. This difficulty, however, does not detract from the importance of these facts and there are among them many observations that speak loud and clear. Consider, for example, the fact pointed out by Susan Carey that many six-year-olds have a vocabulary of 14,000 words or more. As Carey noted, this means that the child must be learning words at a rate of about one an hour for every waking hour of early childhood. To make this rate of learning at all plausible, it must be assumed that much of the vocabulary is acquired after an extremely small number of exposures, perhaps no more than one or two. Where this leaves the complicated reinforcement schedules and elaborate learning strategies that make up such a large part of the literature on learning is a question that surely bears looking into seriously.

While all these studies speak to the fundamental aim of our search to gain a better understanding of the knowledge that native speakers have of their language, the extremely compressed fashion with which of necessity I had to deal with the different contributions of my colleagues has not permitted me to convey to you as clearly as I should like to the nature of the object that we are studying and the extent to which we have a grasp of it. In my experience the best way of providing this kind of insight is by examining a number of real examples in detail. And that is what I propose to do next. The examples that I shall be discussing are taken from the phonic aspect of language. I have chosen these because, on the one hand, this is the facet of language that I am most familiar with, and, on the other hand, because this allows me to bring in Alexander Graham Bell and his contribution to the study of language.

I want to begin by observing that the native speaker of a language knows a great deal about his language that he was never taught. As an example of this untaught knowledge, a list appears below of a number of words chosen from different languages including English. In order to make this a fair test, the English words in my list are words that are unlikely to be familiar to the general public, including most crossword puzzle fans.

ptak thole hlad plast snam mgla vlas flitch dnom rtut

If I now were to ask for a show of hands on each of these 10 words as to whether or not it is to be found in the unabridged Webster's, I am reasonably sure that the majority would vote that *thole, plast* and *flitch* are English words, whereas the rest are not English. This evidently gives rise to a question: Since you have never seen any of the words in the list, how do you know that some are English and others are not? The answer is words judged not English have letter sequences not found in English. This implies that in learning the words of English the normal speaker acquires also knowledge about the structure of the words. The curious thing about this knowledge is that it is acquired although it is never taught, for English-speaking parents—I can swear that this is true of my wife and me as well as some of our acquaintances—do not normally draw their children's attention to the fact that consonant sequences that begin English words are subject to certain restrictions which exclude words such as *ptak, snam* and *rtut,* but allow *thole, flitch* and *plast.* Nonetheless, in the absence of any overt teaching, speakers acquire this knowledge somehow, and this is surely a puzzle worthy of the attention of some learning theorist.

In order to get some insight into how humans acquire knowledge about their language without being taught, it is necessary to understand the character of the knowledge that is being acquired. It is obvious that in the example under discussion that knowledge being acquired concerns the sounds and sound sequences found in English.

Linguists have special ways of dealing with sounds which, incidentally, derive in part from the work of Alexander Graham Bell and that of his father A. Melville Bell. We turn, therefore, at this point to a discussion of the Bells' contribution to the science of language.

As is well known, Alexander Graham Bell was a speech therapist by profession. His specialty was the teaching of speech to the deaf, and according to all reports he was an extraordinarily gifted and successful practitioner of this difficult art. Speech therapy was the profession of many members of the Bell family, as shown on the bottom of the accompanying advertisement, which A. Melville Bell included at the end of his book *Visible Speech*. Speech therapy was a sort of family enterprise which the head of the family practiced in London and other members in other parts of Great Britain. What differentiated A. Melville Bell from most speech therapists was that he was interested not only in the practical aspects of his work but also in its scientific foundations. As we shall see, he involved his son in this work, the future inventor of the telephone, and on one issue of importance the latter made a contribution that went far beyond that of his father.

A. Melville Bell's analysis of spoken language proceeds from the observation that the production of speech sounds involves the coordinated activity of a number of different organs such as the lips, the tongue, the velum, and the larynx, which together make up what traditionally has been called the human vocal tract. From this point of view, the act of speaking is an elaborate gymnastics or choreography executed by different speech organs. In the book *Visible Speech*, we find a systematic account of the different activities that each speech organ is capable of, together with a discus-

PROFESSIONAL CARD.

Mr A. MELVILLE BELL, Author of 'Visible Speech,' may be consulted in all Cases of Impediment or Defect of Speech, Vocal Weakness, Monotony, Oratorical Ineffectiveness, &c.

STAMMERING AND STUTTERING.

The experience of upwards of Twenty-five years' Practice enables Mr A. MELVILLE BELL to undertake the permanent, and, in most cases, the speedy Removal of Stammering and other forms of Vocal Impediment.

References of the highest class are furnished to inquirers.

A limited number of Pupils can be accommodated as Boarders; but residence in the Establishment is not required in order to effect a Cure.

DEFECTS OF ARTICULATION.

In cases of Lisping, Burring, and other Single Elementary Defects, the entire Removal of the Faulty Habit rarely needs more than from Six to Twelve Lessons.

Children who are backward in acquiring the power of Speech are trained to the perfect use of their Vocal Organs. Parents or Governesses are invited to be present at the Lessons, and are directed in the means of carrying on the improvement, which is always rapidly commenced.

ELOCUTION.—PRONUNCIATION, READING, DELIVERY, AND ACTION.

Clergymen, Barristers, Members of Parliament, and other Public Readers and Speakers, are Privately Instructed in the Principles and Practice of Effective Delivery, Oratorical Composition, &c.

Ladies and Non-professional Pupils, receive Special Lessons in the art of Reading, &c., according to individual requirements.

VISIBLE SPEECH.—UNIVERSAL ALPHABETICS.

Pupils are practically initiated in the Physiology of Speech, and in the use of the Universal Alphabet, so as to be enabled to produce, and to record, all varieties of Native or Foreign Sounds.

Dialectic peculiarities are corrected; and Foreigners are taught to pronounce English with the characteristics of vernacular utterance;

TERMS.

Single Lessons in any Department, - - - - - -	-	One Guinea.
Cure of Stammering, Stuttering, &c., -	(Twelve Lessons,) -	Ten Guineas.
Removal of Lisping, Burring, &c., - -	(Six Lessons,) -	Four Guineas.
Elocution, Reading, Delivery, &c., -	(Six Lessons,) -	Three Guineas.
Visible Speech.—Vocal Physiology, &c., -	(Six Lessons,) -	Three Guineas.

The following additional Establishments for the Cure of Stammering and for Elocutionary Instruction are conducted (in Edinburgh) by Mr MELVILLE J. BELL; and (in Dublin) by Mr D. C. BELL.

EDINBURGH: No. 13 South Charlotte Street.

DUBLIN: No. 1 Kildare Place.

LONDON N.W., No. 18 HARRINGTON SQUARE,
(Near Regent's Park.)

sion of the different speech sounds that result from particular combinations of activities of specific speech organs.

Consider from this point of view the speech sounds that are produced by blowing air through a narrow opening as found in the words

veal   zeal   sheep   wheel   what

Sounds produced in this fashion are called *continuants*. One of the things that differentiate one continuant from another is the organ or organs actively involved in its formation, especially the constrictions—that is, the places in the vocal tract that are maximally narrowed when the sound in question is produced—and the organs actively effecting the narrowing, as shown

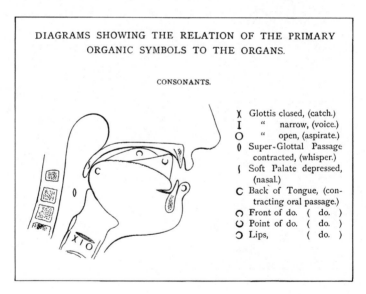

DIAGRAMS SHOWING THE RELATION OF THE PRIMARY
ORGANIC SYMBOLS TO THE ORGANS.

CONSONANTS.

Χ Glottis closed, (catch.)
I   "   narrow, (voice.)
O   "   open, (aspirate.)
() Super-Glottal Passage
     contracted, (whisper.)
(Soft Palate depressed,
     (nasal.)
C Back of Tongue, (con-
     tracting oral passage.)
○ Front of do.   (do.)
U Point of do.   (do.)
Ɔ Lips,   (do.)

in the drawing, which is reproduced from *Visible Speech*. Bell distinguished basically four constrictions: in /f/ * the constriction is formed by raising the lower lip; in /z/ and /s/ the constriction is formed by raising the blade of the tongue, whereas in /$ç^w$/ and /x^w/ there are two constrictions, one formed with the lower lip and the other with the tongue body, or *dorsum*. A further mechanism that is involved in distinguishing one sound from another is whether or not the sound is produced with the accompaniment of vocal cord vibration: /z v/ are; /s $ç^w$ x^w/ are not. This fact can readily be verified by placing one's fingertips on the large (thyroid) cartilage in the front of the neck and pronouncing the sounds in question. When the vocal cords vibrate, this can be detected by a slight throbbing sensation in the fingertips. Finally, for purposes of this discussion, we need to identify one additional mechanism. It is the mechanism that produces strident sounds, such as /f v s z č ǰ/ and distinguishes them from the rest. It consists in directing the air stream against the sharp edges of the upper teeth, thereby producing audible turbulence.

We have thus identified five distinct mechanisms that are involved in the production of the continuant sounds under discussion. We label these for present purposes as follows:

- The raising of the lower lip—labial
- The raising of the tongue blade—coronal
- The raising of the tongue body—dorsal
- Vocal vibration—voicing
- Air stream directed at upper teeth—strident

When two or more mechanisms are activated the perceptual effect is that of a single sound. Thus, both /z/

---

* Slash marks mean the sound represented by the letter.

as in *zeal* and /s/ as is *seal* are perceived as single sounds, although in the production of /z/ one more mechanism (voicing) is activated than in the production of /s/. As shown in the drawing, the *Visible Speech* alphabet had a special symbol to represent each of these mechanisms,—for example, the labial mechanism is represented there by a semi-circle open to the left, the coronal mechanism by a semi-circle open to the top, etc. When two or more mechanisms are activated in the production of a given sound the symbolic representation becomes rather cumbersome. It is therefore more convenient to represent the same information by means of a matrix such as the one in the accompanying chart (on page 51).

The claim made explicitly by A. Melville Bell in his *Visible Speech* is that he had identified all mechanisms that are relevant in the production of sounds in any spoken language. If this claim is correct, then it should be possible for an appropriately trained person to analyze any sound whatever in terms of the mechanisms involved in its production, especially since the number of mechanisms is fairly small. Moreover, it should also be possible for a trained person to produce sounds represented in this notation that he had never heard before. That is exactly how Bell saw the matter and he set about demonstrating it in a most dramatic fashion. I quote from *Visible Speech* (p. 22) :

For the sake of showing the mode in which the experiments were conducted, the following description is quoted from a letter to the "Reader," by Alexander J. Ellis, Esq., F.R.S. :—

The mode of procedure was as follows : Mr. Bell sent his two Sons, who were to read the writing, out of the room,—it is interesting to know that the elder, who read all the words in this case, had only five

weeks' instruction in the use of the Alphabet,—and
I dictated slowly and distinctly the sounds which I

		LABIAL	CORONAL	DORSAL	VOICED	STRIDENT
f	*feel*	+	−	−	−	+
v	*veal*	+	−	−	+	+
xʷ	*what*	+	−	+	−	−
s	*seal*	−	+	−	−	+
z	*zeal*	−	+	−	+	+
š	*she'll*	−	+	−	−	+
ž	*rouge*	−	+	−	+	+
č	*cheap*	−	+	−	−	+
ǰ	*jeep*	−	+	−	+	+
x	*Bach*	−	−	+	−	−
p	*peal*	+	−	−	−	−
d	*deal*	−	+	−	+	−
k	*keel*	−	−	+	−	−

wished to be written. These consisted of a few words in Latin, pronounced first as at Eton, then as in Italy, and then according to some theoretical notions of how Latins might have uttered them. Then came some English provincialisms and affected pronunciations; the words 'how odd' being given in several distinct ways. Suddenly German provincialisms were introduced. Then discriminations of sounds often confused. . . . Some Arabic, some Cockney-English, with an introduced Arabic guttural, some mispronounced Spanish, and a variety of vowels and diphthongs. . . . The result was perfectly satisfactory: —that is, Mr. Bell wrote down my queer and purposely-exaggerated pronunciations and mispronunciations, and delicate distinctions, in such a manner that his Sons, not having heard them, so uttered them as to surprise me by the extremely correct echo of my own voice. . . . Accent, tone drawl, brevity, indistinctness, were all reproduced with surprising accuracy. Being on the watch, I could, as it were, trace the alphabet in the lips of the readers. I think, then, that Mr. Bell is justified in the somewhat bold title which he has assumed for his mode of writing—"Visible Speech."

The quaintness of this testimonial should not be permitted to obscure the serious point that Bell attempted to establish by means of his demonstration, namely, that all sounds of all known languages can be produced given the very restricted information about a small number of mechanisms that is provided by *Visible Speech*. Anybody who controls all the mechanisms singly and in combination can produce any speech sound whatever. It is, therefore, these mechanisms and not the individual sounds of language that are the fundamental building blocks of speech. This insight, which in the last quarter-century has become almost a truism among students of language, was stated explicitly in the early 1900's by Alexander Graham Bell in a series of lectures that he delivered to the American Association to Promote the Teaching of Speech to the Deaf. (It should be noted that Bell's terms *constriction* and *position* are synonymous with what has been termed *mechanism* here.)

> What we term an element of speech may in reality . . . be a combination of positions. The true element of articulation, I think, is a constriction or position of the vocal organs rather than a sound. Combinations of positions yield new sounds, just as combinations of chemical elements yield new substances. Water is a substance of very different character from either of the gases of which it is formed; and the vowel *oo* is a sound of very different character from that of any of its elementary positions.
> When we symbolize positions, the organic relations of speech sounds to one another can be shown by means of an equation; for example
> English $wh = P + P'$ *
> German $ch = P'$
> hence German ch = English $wh - P$
> The equation asserts that the English wh without labial constriction is the German ch. [*The Mechanism of Speech*, pp. 38–39]

I remarked above that during the last quarter-century it had become almost a truism among students of language that the elementary building blocks of lan-

---

* The symbol P in Bell's usage represents the phonetic feature labiality and P' represents the phonetic feature dorsality.

guage are not the sounds but the mechanisms or features as they are now called. While this idea has obvious plausibility as far as the physical production of speech is concerned, it is by no means self-evident that it is useful for other aspects of language. I shall not attempt to support this proposition directly. I shall deal with the data in one case as if the proposition were true and compare it to a solution where this assumption is not made. It will turn out that the obviously superior solution is that one couched in terms of features rather than in terms of speech sounds. I take this as partial evidence in support of the view that features rather than sounds are the basic elements of language in all its manifestations.

1. a) bus bush batch buzz garage judge
   b) cup cut cake cough sixth
   c) cab cad cog cam can song sea shoe sill flower

If you say to yourself the plural forms of the words in (1) you will notice that there is not one but three plural suffixes in English, one for each of the three separate sets of words in (1). We add an extra syllable /iz/ in forming the plural of the words in (1a) ; we add /s/ for the plural of the words in (1b), and we add /z/ to form the plural of the words in (1c). One can readily show that it is not the case that we memorize the plural form of every word we learn, for we know how to form the plurals of words we have never encountered before. Specifically, think of the plurals of the three English words mentioned earlier:

flitch   plast   thole

I am sure that most people here would agree that they know the plural forms of these previously unheard words and that these are respectively

> flitches (like busses [1a])
> plasts (like cups [1b])
> tholes (like cabs [1c])

This forces us to conclude that speakers of English know a rule for the formation of plurals of nouns in spite of the fact that they have not been taught this rule by their parents.

It is necessary to be clear about the status of a rule such as the plural rule under discussion here. They are part of the knowledge that English speakers have and that people who do not know English normally do not possess. Knowing the rule that determines the phonetic actualization of the plural in English is, therefore, much like knowing that the device whose invention we commemorate in 1976 is called *telephone* rather than *farspeaker* (*cf. loudspeaker*), *phonex* or *glub*. The main difference between knowing the rule for the plural and knowing the word *telephone* is that the latter is conscious knowledge about which the speaker can answer direct questions, whereas knowledge of the plural rule and similar matters is largely unconscious and conceivably might never be accessible to consciousness. This fact, it should be noted at once, does not render such knowledge inaccessible to psychologists or linguists—that is, to scientists whose subject of inquiry is the speaker and his knowledge. Tacit knowledge can be established by the same methods that were used to establish other things inaccessible to direct observation such as the nature of the chemical bond or the structure of the gene.

The question that we want to answer is, In what form does the English speaker internalize his knowledge of the plural rule? An obvious candidate is:

2)

    a) If the noun ends with /s z š ž č ǰ/, add /iz/.

b) Otherwise, if the noun ends with /p t k f θ/, add /s/.

c) Otherwise, add /z/.

It is important to note about this rule that it is formulated in terms of speech sounds rather than in terms of mechanisms or features. In the light of the above discussion which suggested that features rather than sounds are the ultimate constituents of language, an attempt might be made to reformulate the rule in terms of features. The first move that one might make might be to replace each of the alphabetic symbols in (b) by its feature composition as shown in the preceding chart. Specifically, this means that one might replace /s/ by the feature complex /nonlabial-coronal-nondorsal-nonvoiced-strident/ /z/ by the same set of features except that in place of nonvoiced it would contain the feature voiced, etc. It is not easy to see where such a translation of the rule into feature terminology gets us. In fact, it gets us nowhere until we observe that with a chart such as that given earlier, it is possible to designate groups of sounds by mentioning one or two features. Thus, for example, if we asked for all and only sounds that are labial we would get the group /f v x^w p/; whereas if we asked for the sounds that are strident we would get /f v s z s z č ǰ/. Suppose now that we were to utilize this idea in the formulation of the plural rule and characterize each of the different lists of sounds by the minimum number of features that suffice to designate the group unambiguously. We should then get in place of (2),

3)
a) If the noun ends with a sound that is /coronal-strident/, add /iz/.

b) Otherwise, if the noun ends with a sound that is /nonvoiced/, add /s/.
c) Otherwise, add /z/.

Having formulated an alternative to the rule given above as (2), our task now is to determine which of the two alternatives is the one that English speakers use. The test that we shall use is one suggested to me some years ago by Lise Menn. It consists of asking English speakers to form the plural of a foreign word which ends with a sound that does not occur in English. A good example, Ms. Menn suggested, is the German name *Bach,* as in Johann Sebastian. If English speakers were operating in accordance with rule (2), they would have to reject options (a) and (b) and form the plural in accordance with option (c); that is, they would say that the plural of /bax/ is /baxz/ with a word final /z/. If, on the other hand, English speakers were operating in accordance with rule (3), they would have to perform a feature analysis of /x/ which would tell them that the sound is /nonlabial-noncoronal-dorsal-nonvoiced-nonstrident/. Given this feature composition, the plural of /bax/ could not be formed in accordance with option (a), since /x/ is neither coronal nor strident; it would, however, have to be formed in accordance with option (b) since /x/ is /nonvoiced/. In other words, if speakers operated in conformity with rule (3), their output would be /baxs/, which, as is perfectly obvious, is also the response that the majority of English speakers would make. We must, therefore, conclude that the formulation (3) of the plural rule in terms of features, and not the formulation (2) in terms of speech sounds, correctly represents the knowledge of English speakers.

There is yet another, more important inference to

be drawn from the fact that English speakers can apply the plural rule to a word ending with a sound that is not part of the repertory of English. In order to apply the rule, the speaker has to be able to establish that the foreign sound in question is nonvoiced. He must therefore have knowledge that allows him to determine the phonetic mechanism involved in the production of a sound that is not part of his language. The curious thing about such knowledge is that not only is there no indication that it might ever have been taught to speakers, there is also no indication that speakers could ever have acquired such knowledge. Think what evidence would have to be marshaled to support the claim that the knowledge in question was acquired. One would have to point to experiences in the life of the average English speaker that would permit him to acquire knowledge that is otherwise possessed only by phoneticians who have undergone rigorous training of the type Alexander Graham Bell received from his father. As this is obviously implausible, one is led to contemplate the possibility that at least some knowledge available to speakers is innate. In fact, there appears to be a certain amount of independent evidence that knowledge of the feature composition of sounds is available to children long before they could possibly have learned a language. Experiments conducted by Peter Eimas at Brown University and by Earl Butterfield at the University of Kansas have established that the ability to discriminate voiced from nonvoiced speech sounds is present in children practically at birth, but this ability presupposes knowledge which allows persons to determine the feature composition of speech sounds, that is, the same knowledge that is required in order to account for the ability of English speakers to form plurals of words with non-English sounds. The suggestion that this type of knowledge might be innate is, therefore, far from implausible.

This brings me to the end of what I have to say about the knowledge that speakers have of their language. What remains for me to do is to indicate how the information we have just reviewed helps us in trying to understand manifestations of the human cognitive capacity in domains other than language, how it might help us understand the human capacity to draw inferences, perform computations, play games with elaborate rules, interact with one another and uncover significant truths about the nature of the world around us and within us. If these manifestations of man's mind are at all like language, then we must expect to find that large portions of the knowledge on which they are based will be inaccessible to consciousness, that some of this knowledge will be innate, and that only a modest fraction of the total will have been acquired as the result of overt teaching. I must confess that I felt somewhat uneasy when I noticed that I was drawing attention to the possibility that teaching might play only a marginal role in the acquisition of knowledge. I was concerned about the effects that this remark might have on next year's enrollments in the M.I.T. linguistics program. But then I recalled the effect that the Surgeon General's warning printed on every pack of cigarettes has had on the popularity of smoking in this country and concluded that there was nothing to worry about since very little indeed is learned as the result of direct instruction.

FUTURE IMPACT OF COMPUTERS AND INFORMATION PROCESSING

Michael L. Dertouzos

Professor Dertouzos was born in Athens, Greece, and came to the United States for undergraduate studies under a Fulbright Fellowship. He received his bachelor's degree in Electrical Engineering at the University of Arkansas and his Ph.D. in Electrical Engineering in 1964 at M.I.T. He joined the faculty of that department as Assistant Professor. He was appointed Associate Professor in 1968 and Professor in 1973. He is now Director of the Laboratory for Computer Science, which was known previously as Project MAC. Professor Dertouzos entered Project MAC as a doctoral student when the laboratory was founded. His early research interests were in the area of thresh- 👉

old logic and on-line circuit design. He has written many books and papers and is a holder of patents in computers and electrical engineering. He was awarded the B.J. Thompson award of the Institute of Electrical and Electronic Engineers for the best paper in IEEE publications in 1968. He has a great deal of interest in undergraduate education and it was he who introduced digital systems into the M.I.T. core curriculum and later on developed the first professional course in computer science for the core curriculum at M.I.T. He today calls himself a specialist in control robotics, the use of computers for the control of physical processes and in programming languages.

This is an interim report on a year-long study we are conducting on the future impact of computers and information processing. The study was motivated by several serious questions which confronted us as we changed the name of our laboratory from "Project MAC" to the "M.I.T. Laboratory for Computer Science," implying a change in direction from a single project to a broad range of research activities in computer science. In broadening our outlook we had to sharpen our priorities so that we could apply our resources to problems which are indeed critical in advancing the field. It was this challenge which led us to undertake the study.

There is no question that we are in the middle of an information revolution and that the computer promises to affect our life in major ways. After a quarter-century of existence, the computer field seems to be reaching a certain maturity, at least in terms of our ability to forecast many future hardware and some future software developments. Many important questions, however, remain unanswered: For example, technological developments in the next 10 to 20 years might make possible automation of the dispensation of services and the production of goods, understanding of natural language, new ways for educating and entertaining mankind, new ways of communicating, conducting business, and governing, and, in general, the assumption by the computer of progressively higher levels of what are now considered human skills, knowledge and intelligence. There is, of course, considerable debate as to whether such feats can be achieved strictly on technological grounds. There also is considerable debate on the relationship between such technological progress and society's need for or reaction to its onset.

To begin addressing these questions, we formed a steering committee composed of Daniel Bell from Harvard, John McCarthy from Stanford, and Victor Vyssotsky from Bell Laboratories, together with Joel Moses and myself from M.I.T., who are acting as co-directors of the study. The committee soon concluded that the nature of the questions did not admit of simple answers and agreed that the purpose of the study should be to raise technical and societal issues, suggest approaches and provide guidance to researchers, decision makers and the interested public at large. We then proceeded to identify some 20 specific topics which would form the substance of the study, and which would deal with these questions in the time frame of the next 10 to 20 years. The next step was to identify individuals who would write papers on each of these topics. To this end, we did not seek generalists who could supply broad surveys but, rather, individuals whose experience, performance record, and point of view qualified them to deal comprehensively with important aspects of each of these topics. The topics and authors given in the list on pages 66–67 show that we have succeeded in attracting a remarkably qualified group of individuals to participate in this look at the future. Indeed, many of these individuals are in a position to affect the outcome of their prophecies.

Our study is broadly divided into three categories. The first category deals with the possible directions in which technology may move. It addresses questions such as "What are the technological frontiers?," "What are the perceived opportunities?," "What are the foreseeable limits?" The second part of the study has to do with uses—uses in the home, uses in government, uses in automation, and so on. The third category has to do with the social impact of these technical developments, ranging from the effect upon our individual lives to effects upon our organizations and institutions.

The form of each paper was left to the discretion of the author, consistent with the overall purpose and target audience of the study. It was understood, however, that each paper would examine the current frontiers and focus on extrapolation of visible trends as well as on the author's point of view as to the directions in which future developments might or should evolve. The deadline for drafts of the papers was March 1976, and all papers are being circulated to all authors for mutual critiques. In addition, we have two official critics participating in the study, whose sole purpose is to provide thoughtful criticism on all the essays. The objective of this rather extensive internal review process is not to force compromise between authors and critics but, rather, to sharpen up the positions taken in the papers and to eliminate the typical misunderstandings that develop between critic and author when they operate independently.

Since, as I have indicated, we have just completed the drafting stage of the study, the results presented at the convocation must be considered as preliminary in nature. Accordingly, the finished study, which will be published in book form in the middle of 1977, may differ in substantial ways from the work which I sketch out here. Further, I can do little more than touch on the highlights of our discussions and thus am unable to do justice to some of the more fundamental and detailed discussions which occurred during our sessions. Finally, before summarizing our discussions, I would like to take note of the fact that this study is being sponsored by A.T.&T., I.B.M., the Office of Naval Research and M.I.T.

STUDY OF THE FUTURE IMPACT OF COMPUTERS
AND INFORMATION PROCESSING
Titles and Authors

Introduction:	M. L. Dertouzos, M.I.T.
	J. Moses, M.I.T.
Background:	
Historic Review of Computer Developments	M. Denicoff, Office of Naval Research
Hardware Prospects and Limitations	R. N. Noyce, INTEL Corporation
Technological frontiers:	
Computer Science and the Representation of Knowledge	M. L. Minsky, M.I.T.
Current Research Frontiers in Computer Science	A. J. Perlis, Yale University
The Peoplization of Computers	T. Winograd, Stanford University
Some future applications:	
Computer-Communication Services for the Home	J. Moses, M.I.T.
Computers and Modeling	M. Shubik, Yale University
Computers and Education	S. A. Papert, M.I.T.
Computers and Government	J. C. R. Licklider, M.I.T.

Computers and Communications	B. O. Evans, IBM Corporation
The Return of the Sunday Painter or the Computer and the Visual Arts	N. Negroponte, M.I.T.
Use of Computers for Business Functions	V. A. Vyssotsky, Bell Telephone Laboratories
Scientific Use of Computers	S. Fernbach, Lawrence Livermore Laboratories
Computers and Automation	M. L. Dertouzos, M.I.T.

Impact on society:

The Social Framework	D. Bell, Harvard University
Economics of Information	K. J. Arrow, Harvard University
The Consequences of Computers for Centralization and Decentralization	H. Simon, Carnegie-Mellon University
Social and Regulatory Framework of Computers and Communications	R. Noll, California Institute of Technology
The Computer and World Affairs	R. G. Gilpin, Princeton University

Official critics:

| | J. McCarthy, Stanford University |
| | J. Weizenbaum, M.I.T. |

In assessing the future directions of technology, it is necessary first to look at the basic electronic components out of which the machine hardware is built. The complexity of integrated circuits has been increasing at a spectacular rate since 1960 as a result of increasing circuit density, increasing size of integrated circuit chips, and increasing functional density through circuit innovations.

The main motivation has been economic and has resulted in a decrease of cost by a factor of about two each year, since 1960. This cost reduction has come primarily from reduction of the cost of interconnection of components, resulting from the fact that in integrated circuit technology the interconnections are created simultaneously with the components. The projections made by the specialists in this area suggest that by 1990 we may be able to obtain further cost reductions by a factor in the range of 100 to 1000, depending on whether one is pessimistic or optimistic. The fundamental physical limits, such as quantum effects, are several orders of magnitude away. Even when these limits are reached, they might be pushed further away by the use of cryogenics or other techniques.

Among the more immediate factors which might limit the pace of increasing complexity in integrated components are the need for better than one-micron resolution by photolithography, further innovation in circuit forms, and the development of mass markets to justify increasing circuit design costs. Given the fact that semi-conductor component manufacturers currently are investing about 10 percent of their sales income in research and development expenditures, it is apparent that the point of diminishing returns of these investments has not yet been reached. Within this overall context, it is possible that by 1980 the bits of memory in a small chip will have increased from today's level of 10,000 to about 100,000. By 1990 this density may increase by another factor of 10. For logic gates and components used in processors, the density is expected to lag that of memory chips by a factor of three to 10 as a result of interconnection layout problems.

What developments do we expect in machines made from these increasingly complex components? We can gain some insight by looking at what has been the history of computer usage over the past 20 years in an organization such as the Energy Research and Development Administration, which currently has a total of some 2,000 computers in operation. Since the introduction in the mid-1960's of small machines, costing less than $50,000 (minicomputers), there has been a leveling out of the growth of large machines, costing more than $1,000,000, and of medium machines. At the same time, the number of small machines has continued to grow dramatically, until today they constitute about 1,400 out of the 2,000 computers in ERDA. With regard to future trends, it is possible that this pattern of growth is conservative, because we now have microprocessors which cost considerably less than $50,000, some as cheap as $10 each. These machines are now starting to find their way into research installations and institutes as well as into the design of products. Thus, one expectation of the future is a dramatic growth in the use of very small machines. There also is consensus among our participants that we will see more parallelism, that is, the simultaneous use of several processors to solve one problem; the most commonly cited example of such use is that of weather prediction where, currently, single large computers are taking some 14 hours to work out a detailed 24-hour forecast. The lower cost components also bring within the range of possibility a home computer of remarkable power, comparable to what would have been called a

MICHAEL L. DERTOUZOS

medium machine some years ago, yet costing only as
much as a color television set. It is also agreed that
there will be substantial developments in input-output
devices. Within the time frame of the study it is expected that we will be able to communicate more easily
and effectively with computers through terminals
which have much better visual displays as well as some
form of oral communication capability. There also will
be increasing developments of specialized terminals
which have high bandwidth input-output capacity and,
also, increasing networking of computers for accessing
and processing information. In addition, the decreasing
costs of microprocessors will result in their increasing
use to control various kinds of industrial and consumer
processes and products such as, for example, the combustion of automobile engines.

Continuing our assessment of the probable trends of
computer technology, I turn now to what we might put
upon these machines, that is, the software. I use the
term in a very generic sense to encompass not just languages and basic operating systems but also the more
advanced services that these machines might perform.
There is a clear consensus among our participants that
software costs are high and rising, partly because of
the difficulty of designing efficient programs for the increasingly powerful machines, and partly because of
the difficulty of finding mistakes and "debugging" these
complex programs. Some see the increase in complexity
and proliferation of hardware as just an enlargement
of the base over which costly software mistakes can be
made. There is, accordingly, a great need for ways that
circumvent the problems of complex software production. Automatic programming, the ability to generate
new programs by use of existing programs, is one avenue of development being pursued. Another direction
involves the development of more intelligent programs

—programs that, within a restricted context, understand what one wishes to accomplish. Another direction being pursued is what one participant calls the "peoplization" of computers, that is, the development of machines, input-output devices and software that responds to varying personal styles of users. This approach is to be contrasted with the current relatively inflexible machine modes to which all users must conform—a process which has been called "computerization" of people. Within this context one can view the evolution of attitudes toward computers along three identifiable generations. The first generation attitude views computers as giant calculators that could manipulate numbers and solve numerical problems. The second generation attitude views the computer as a data processor for manipulating large data bases and symbols, such as names and payrolls. The developing third generation attitude is to consider computers as *knowledge processors* which can work with structured bodies of knowledge and carry out intelligent reasoning processes and interactions.

These being the trends in the development of the underlying computer technology, let us now take a look at some of the anticipated uses of computers.

There was much discussion in our sessions about the use of computers in the home. As I listened to the description of the early speculations about uses of the telephone in the home, I was reminded of our own speculations about computers in the home—they sound very similar. Three modes of computer use in the home are envisioned. First, an increasing use of microprocessors to control and operate appliances. Second, a "stand alone" computer which would be used for recreation, education and computing tasks within the home. Some see the computer's ability to offer more active and participatory forms of entertainment as an alternative to the current, relatively passive entertainment offered by television. Some early modes of such recreation are already available in the market in the form of electronic Ping-Pong games which use the television screen as a display device. Games of much greater sophistication are possible as we know from the complicated space-war and other games developed by our students here at M.I.T. For example, one can imagine a chess game which can be tuned to have different and controllable levels of skill to suit the personality and skill of the human player. Another important use of the home computer is for education. I will discuss the educational use of computers in more detail later and will only remark here that the emphasis will be on learning by doing rather than on the accumulation of information through routine drill and practice. The third mode of anticipated computer use in the home arises when the machine is connected to an outside network capable of supplying information and certain services. The much discussed possibility of having access to vast library resources is not considered as a realistic possibility during the time frame of the study, because of the mix of technical, economic and organizational problems surrounding this issue. Nevertheless, the technology seems to make possible the use of home terminals for obtaining facts, for obtaining advice and for receiving book information in specified domains from some central location or distributed locations.

Turning to another area, much of our mail today consists of financial and routine information which could equally well be delivered electronically. Considering the deteriorating quality of our current mail service and the early success of electronic message systems, there is consensus that we will see an increasing use of electronic mail in the next two decades. Another anticipated use of the home computer is what has been called reverse advertising or catalogue shopping via computer. For example, one could say to the terminal, "I would like a boat about 16 feet long with such and such characteristics, costing between $2000 and $3000. What have you got?" Information would then flow back to the consumer from a variety of manufacturers rather than from a single manufacturer's catalogue, thus expanding the realm of individual choice.

News filtering is another issue discussed in relation to the computer in the home. It is widely recognized by all our participants that one of the major problems that we face is the explosion in unstructured or raw information which is being continuously created. Some of this explosion is no doubt due to the existence of computers, but a more basic cause seems to be the explosion in quantity and variety of economic, political and intellectual activity throughout the world. We are thus bombarded by huge amounts of information and feel the need for techniques that will improve its quality and that will make possible its digestion in an easier way. News filtering would be one such approach. By incorporating in your home computer a small program that will identify news stories of interest to you —for example, having to do with certain sports or development of news in a certain part of the world, whatever your interests might be—you could arrange for your home terminal to print out only a restricted set of news that might come over a news service wire to which your terminal had access. Observe that while there was no consensus among our participants that the demands for home uses of computers would bring forth the necessary large capital and software investment, it was agreed that work on these questions should be part of the research and development agenda over the coming decade.

So far, most uses of the computer for education have

been within the context of existing educational theories and practices. Thus, a good deal of work has been done on the equivalent of drill and practice education, namely, replacement of a teacher with a more patient, if less flexible, computer. Of course there have been innovative and effective uses outside this traditional framework, such as the use of management games that confront students with very realistic situations in which they must make decisions in real time and thus get some measure of learning by experience.

Only in the last few years have there been attempts to use the computer to create wholly new learning environments and experiences for young children. In contrast to the standard view of a computer as a machine acting *on* learners, the thrust of these new developments emphasizes the spontaneous activity of the child, the self-organizing aspects of the learning process: in short, the Piagetian view of the child as an intellectual agent who reconstructs the essential components of the mind in response to incoming information and interactions with the environment.

Currently, various directions of research are being pursued to create situations in mathematics, music and in physics where, through use of very simple programming, the child can direct a computer to perform operations which achieve objectives set by the child. Invariably, the first attempts at such programming are unsuccessful, and the basic learning really comes as the child learns to "debug" the program until the desired objective has been accomplished. Information about ideas or about the physical world gathered by this kind of experience becomes part of the working mental equipment of the child and can indeed create wholly new attitudes toward learning.

As one example of this approach I would like to describe a learning situation in which a child experimentally determines the nature of a circle. Connected to a computer is a device—called a "turtle" because of its general shape—which is motorized to travel in any direction about the floor in response to appropriate instructions from a computer and to leave a mark of the path that it has followed. The child is taught a program which allows him to give two kinds of instructions to the turtle: first, he can instruct the turtle to move forward a specified distance along the direction in which the turtle currently is oriented; second, he can instruct the turtle to rotate about its center so as to change its orientation by a specified angle. Armed with this programming ability, the child is given the task of programming the computer so as to draw a circle on the floor. Ultimately, almost all children discover that the program which works is one which instructs the turtle to move forward a given distance, turn through a particular angle, move ahead the same given distance, turn again through the particular angle, and so on until the turtle arrives back at the starting point. The child also discovers that the path comes closer and closer to the smooth curve of a circle as the distance and angle increments are made smaller and smaller, leading him to develop intuitively the mathematical concept of curvature. Thus, the potentially most fruitful educational uses of the computer are seen as uses which get children involved in nontrivial projects such as this which lead them to think in new ways about many experiences.

In the area of business functions, there is widespread agreement that the use of paper will continue to decrease because of increasing use of electronic records, record keeping and communications. This trend to the "paperless office" will substantially change certain types of clerical force requirements. It is expected that management information systems, which have not been terribly successful to date, will move away from their current preoccupation with numerically accurate results aimed at top business managers, and will tend more toward providing information to people throughout the organization, for the purpose of helping them discharge their functions more effectively. Quite a few of our participants agree that there will be qualitative changes in the way businesses operate, and the most important among these changes will be computer developments in automation; the structuring of knowledge; and the assumption by computers of what we might call knowledge work, starting at a relatively low level. There is some expectation that these changes will be facilitated by the familiarity of new employees with computers as a result of a high school or college experience. Interestingly enough, the projections are low for improvements in productivity arising from further computerization of business functions. One participant projects improvements of the order of 30 to 40 percent by the turn of the century—so we are not talking about something that will happen overnight.

Automation is another area where current and foreseeable developments in computers and computer science promise to have major effects. While the popular view of an automated future involves a heartless and mechanistic society geared to productivity and efficiency, we have the choice to create the opposite, namely, a future in which computers will be used to improve and humanize our way of life.

As an example, consider what might be involved in the acquisition of a pair of shoes a few years hence. You would walk into the local computerized shoe center and place your foot into a box designed to measure its shape. You would then adjust the pressure of the measuring straps around your foot until the feel is comfortable, and you would then push a button to record the measured shape. The shape of both feet would be encoded on a card, which you then would insert into a style-selection machine along with your credit card. The style machine would display a range of choices many times larger than available in shoe stores today because the shoes would be stored in the form of raw materials rather than finished products.

Having chosen your style, you would push the appropriate buttons to signal your choice and the style-choice machine would collect the price from your credit card account and would then initiate the automatic manufacture of your shoes in an adjoining room under direction of a computer which controls the flow of materials and the operation of the programmable assembly machines which produce your pair of shoes in a matter of minutes.

While this may seem farfetched, all the needed components, with one exception, are here today in at least rudimentary form. The exception is a programmable assembler with visual input which might be needed for inspection or other operations. Whether such a robot will be available in a few years is difficult to forecast, since such a development depends upon research results in automatic retinas and in the use of certain knowledge by machines, as well as on economic questions. Were such a development to take place, it would not only provide a wider choice of products at a lower price, but could also revive the decaying domestic shoe industry, which might disappear from our GNP if current governmental subsidies were removed.

There are many other situations in which automation using computers could be socially or economically beneficial. A few examples are the use of automated robots in hostile environments such as spray painting and mining, the replacement of humans in mindless tasks, or traffic control accomplished by a distributed system with a computer at every city intersection and a computer intercommunication network that follows city arteries. Significantly, along with these social and economic gains, computerized automation offers the possibility of reversing the trend of homogeneous products and services brought upon us by the industrial revolution. Let us not forget that as individuals we are the most variable of demand centers. The computer offers the promise of adapting products and services to our differing needs, thereby restoring some of the benefits of individualization that existed before the industrial revolution while maintaining the cost benefits of automated production.

Let me now turn to our discussion of the broader effects which computers and information processing might have on society over the next 10 to 20 years.

Whereas the forces influencing the development of computer technology in the past 20 years were mainly internal to the technology and its applications, the forces that will influence developments in the next 20 years seem to lie in the interaction between technology and society. Modern communications and computers have moved us from a world in which information was a scarce commodity, to be cherished and preserved, to a world so full of information that what is scarce is the capacity to process and digest information. We are, in fact, confronted with the economics of information, and the social and economic advantages and disadvantages associated with our capacity to use information. In short, information is well on its way toward becoming a dominant commodity.

Some of our participants feel that there probably exists a continuing, long-term trend toward centralization of decision making in both business and government as more and more interdependence is recognized among the matters about which decisions have to be made. It does not appear that computers have contributed substantially to this trend. Some people believe that computers have arrived just in time, because of the powerful new means that they provide for helping decision makers to deal with problems that involve large numbers of interacting variables. One of our colleagues argues the necessity of computer and modern communications systems for the survival of any liberal society which tries to treat individuals as individuals. If society wishes, it can use the computer to open the decision process to inspection, thereby reconciling the notion of central leadership in the setting of goals with the notion of decentralization in the actual decision processes. Further, because of its capacity to compact information and to permit multiple inputs into the decision process from a variety of sources, the computer can allow a more meaningful participation of both experts and laymen in debates on public policy, often permitting entirely independent alternative analyses of the same problem.

We had a particularly interesting discussion on the effects which the computer might have on our planning and decision making in the future through computer modeling of complex problems involving economic, technological and political interactions. There is much concern about how we can improve modeling so that our models are more accurate abstractions of the real world and therefore useful in decision making and planning. One of the difficulties with modeling is that it is much easier to make a large simulation model of a system than it is to judge whether the result is a good model of the system under study, particularly when we are trying to look into the future, as is so often the case. The view was advanced that research on modeling and training of modelers should be a matter of high priority, so that we can make better use of computers in our planning processes at all levels and in all functions of society.

As individual citizens, some of our colleagues expect that they will be affected by the computer more and more in their relations with government, since government is our largest service industry. Heretofore, the applications of computers in the service efforts of government have been concerned mainly with keeping the records straight, but the feeling is that, in the future, computers will be involved increasingly in the execution and delivery of the services.

Some of our colleagues see major influences of the computer in the international sphere. The foremost consequence of this effect is expected to be a continuation of the trend of decreasing international business transaction costs. Electronic communications and computers have created the necessary technological means for the integration of the world economy by multinational corporations and financial institutions, and it is expected that further developments in information processing and computers will lead to increasing interdependence of national economies.

The impact of the computer on weapons and warfare has been profound and its consequences for the nature and conduct of war have still to run their course. On the strategic level, the computer has been a basic factor in the evolution of mutual deterrence which underlines détente. On the tactical level, the computer may well strengthen defense over offense, thus hopefully inhibiting the propensity for nations to resolve their differences through violence. The economic and military changes associated with the impact on international relations of the computer and other advanced technologies suggest that the world may be entering an age where more is gained through economic cooperation and an international division of labor than through strife and conflict. Yet, as one of our colleagues pointed out, this is a belief that has been associated with other great technological innovations and has been proven false. One should not be overly sanguine, therefore, that the computer will deliver us from the scourge of war. Certainly, if properly exploited, the computer and its attendant information-processing applications can give us some new causes for hope.

I would now like to conclude this summary with what I perceive to be common themes at this interim stage of our study. First, there is no doubt that hardware costs are decreasing and that new machines will be developed using microcomputers, parallelism and a great deal of distributed computation. We shall also see distributed data bases and distributed processors and networks. Moreover, communications, especially of the broad-band variety, will continue to be a central and growing factor in the use of computers in the future. Second, we expect that there will be progress in structuring information and knowledge, in improving its quality and in compressing it so that it can be managed. Third, we will hopefully see a decrease of software costs and an adaptation of machines to user needs, with machines serving user functions through more intelligent programs. Fourth, the use of machines interactively through user participation, as we discussed in connection with education, was recognized as a common and powerful trend in other areas of computer use. Finally, as I said at the outset, there is no question that the information revolution is with us. We all have questions about its rate and manner of growth. There is no question, however, that this revolution, together with the foreseeable future developments in computer technology and information processing, will produce sizable and qualitative changes in our way of life. It is our hope that a year hence when we publish our completed study we will be able to be more specific about directions of research and development which might increase the likelihood that these technological and social changes will be beneficial.

COMMUNICATIONS IN THE SECOND CENTURY OF THE TELEPHONE

Arthur C. Clarke

Arthur Clarke is basically an explorer, an explorer of a variety of spaces—outer space, underwater space and the space of our imagination, as the essence of what one young man called "the most" in science fiction. He didn't start out that way. He went to Kings College, graduated with first class honors in physics and mathematics and did a lot of things between then and this career of his. He is among other things the past Chairman of the British Interplanetary Society. He is a member of the Academy of Astronautics of the Royal Astronomical Society and many other scientific ☞

and professional groups. The Franklin Institute's medal testifies to the fact that he originated the idea of the communications satellite in a technical paper published in 1945. In 1962 he was awarded the UNESCO Kalinga Prize for science writing. He has won many other prizes and, in 1969, he shared a nomination for the Oscar with Stanley Kubrick for the screen play of 2001: A Space Odyssey. *In addition, he is the author of some 50 books, including* Imperial Earth, Voice Across the Sea, Profiles of the Future, The Treasure of the Great Reef, *and* Man and Space. *Since the early 1950s, he has been exploring the barrier reef off Australia and Ceylon and has found many ways of putting these explorations into an enormously attractive communicative form.*

Back in 1943, as an extremely callow officer in the Royal Air Force, I was given a mysterious assignment to a fog-shrouded airfield at the southwestern tip of England. It turned out that I was to work with an eccentric group of Americans from something called the Radiation Laboratory of the Massachusetts Institute of Technology. They were led by a bright young physicist name Luis Alvarez, who had invented a radar device that, for a change, did something useful. It could bring down an aircraft in *one* piece, instead of several.

The pioneering struggles of the Ground Control Approach (G.C.A.) "talk-down" system, and our attempts to convince skeptical pilots that we knew exactly where they were (even when we didn't) you'll find in my only *non*-science fiction novel [1] . . . though *Glide Path* certainly would have been science fiction had it appeared at any time before 1940.

I mention this incident from the Paleo-Electronic Era for two reasons. Luis' brainchild provided me with the peaceful environment, totally insulated from all the nasty bombings and invasions happening elsewhere, which allowed me to work out the principles of communications satellites in the spring of 1945. So comsats lie on one of the many roads that lead back to M.I.T.

My second reason for mentioning the primitive technology of a third of a century ago is that it gives a baseline for extrapolation into the future. I can still recall my amazement at the number of vacuum tubes (remember them?) in the G.C.A. Mark 1. It came to the unbelievable total of almost one thousand.

I would have laughed scornfully if some crazy science fiction writer had predicted that one day every engineer would carry at his hip not a Colt .45 but an H/P 45, or its successors—mere handfuls a *dozen times as complex* as our Mark 1. The explosion in complexity, and the implosion in size, are two of the main parameters determining future communications technology.

Now, in discussing that technology, I'm acutely aware that I'm competing with people who have made it their major concern for decades. It's inevitable, therefore, that in the course of this paper I'll invent the wheel not once but several times. My main hope is that *my* wheels will be of novel and interesting shapes, unlike the boringly circular ones produced by all the professionals.

But first, as the commercials say, a few general principles. It's probably true that in communications technology anything that can be conceived, and which does not violate natural laws, can be realized in practice. We may not be able to do it right now, owing to ignorance or economics—but those barriers are liable to be breached with remarkable speed.

For Man is the communicating animal; he demands news, information, entertainment, almost as much as food. In fact, as a functioning human being, he can survive much longer without food—even without water!—than without information, as experiments in sensory deprivation have shown. This is a truly astonishing fact; one could construct a whole philosophy around it. (Don't worry—I won't try.)

So any major advance in communications capability comes into widespread use just as soon as it is practicable. Often sooner; the public can't wait for "state of the art" to settle down. Remember the first clumsy phonographs, radios, tape recorders? And would you believe the date of the first music broadcast? It was barely a year after the invention of the telephone! On

2 April 1877, a "telegraphic harmony" apparatus in Philadelphia sent "Yankee Doodle" to sixteen loud-speakers—well, soft-speakers—in New York's Stein-way Hall. Alexander Graham Bell was in the audience, and one would like to know if he complimented the pro-moter—his now forgotten rival, Elisha Gray, who got to the Patent Office just those fatal few hours too late. . . .[2]

Gray was not the only one to be caught out by the momentum of events. I apologize for quoting the fol-lowing story for the hundredth time, but it is far too relevant to omit, and there may still be a few people around who haven't heard it.

When news of the telephone reached England, pre-sumably a hundred years ago tomorrow through Cyrus Field's cable, the chief engineer of the post office was asked whether this new Yankee invention would be of any practical value. He gave the forthright reply: "No, sir. The Americans have need of the telephone—but we do not. We have plenty of messenger boys."

Before you laugh at this myopic Victorian,[3] please ask yourself this question: Would you, exactly a hun-dred years ago, ever dream that the time would come when this primitive toy would not only be in every home and every office, but would be the essential basis of all social, administrative and business life in the civ-ilized world? Or that one day there would be approxi-mately one instrument for every ten human beings on the planet?

Now, the telephone is a very simple device, which even the nineteenth century could readily mass-pro-duce. In fact, one derivative of the carbon microphone must be near the absolute zero of technological com-plexity. You can make a working—though hardly hi-fi —microphone out of three carpenter's nails, one laid across the other two to form a letter H.

The extraordinary—nay, magical—simplicity of the telephone allowed it to spread over the world with astonishing speed. When we consider the very much more complex devices of the future, is it reasonable to suppose that they too will eventually become features of every home, every office? Well, let me give you another cautionary tale.

In the early 1940's the late John W. Campbell—editor of *Astounding Stories* and undoubtedly the most formidable imagination ever to be flunked at M.I.T.—pooh-poohed the idea of home television. He refused to believe that anything as complex as a TV receiver could ever be made cheap and reliable enough for domestic use.

Public demand certainly disposed of that prophecy. Home TV became available in the Early Neo-Electronic Age—that is, even *before* the solid-state revolution. So let us take it as axiomatic that complexity is no bar to universality. Think of your pocket computers again and march fearlessly into the future . . . trying to imagine the ideal, ultimate communications system—the one that would fulfill all possible fantasies.

Since no holds are barred, what about telepathy? Well, I don't believe in telepathy—but I don't *dis*believe in it either. Certainly some form of electronically assisted mental linkage seems plausible; in fact, this has already been achieved in a very crude form, between men and computers, through monitoring of brain waves. However, I find that *my* mental processes are so incoherent, even when I try to focus and organize them, that I should be very sorry for anyone at the receiving end. Our superhuman successors, if any, may be able to cope; indeed, the development of the right technology might force such an evolutionary advance. Perhaps the best that *we* could manage would be the sharing of emotional states, not the higher intellectual processes. So radio-assisted telepathy might merely lead to some interesting new vices—admittedly, a long-felt want.

Let's stick, therefore, to the recognized sense channels, of which sound and sight are by far the most important. Although one day we will presumably develop transducers for all the senses, just because they are there, I suspect that the law of diminishing returns will set in rather rapidly after the "feelies" and "smellies." These may have some limited applications for entertainment purposes, as anyone who was pulverized by the movie *Earthquake* may agree. (Personally, I'm looking forward to the epic *Nova,* in which the theater's heating system is turned on full blast in the final reel. . . .)

The basic ingredients of the ideal communications device are, therefore, already in common use even today. The standard computer console, with keyboard and visual display, plus hi-fi sound and TV camera, will do very nicely. Through such an instrument (for which I've coined the ugly but perhaps unavoidable name "comsole"—communications console) [4] one could have face-to-face interaction with anyone, anywhere on earth, and send or receive any type of information. I think most of us would settle for this, but there are some other possibilities to consider.

For example: What about *verbal* inputs? Do we really need a keyboard?

I'm sure the answer is "Yes." We want to be able to type out messages, look at them, and edit them before transmission. We need keyboard inputs for privacy, and quietness. A *reliable* voice recognition system, capable of coping with accents, hangovers, ill-fitting dentures [5] and the "human error" that my late friend HAL complained about, represents something many orders of magnitude more complex than a simple alpha-numeric keyboard. It would be a device with capabilities, in a limited area, at least as good as those of a human brain.

Yet assuming that the curves of the last few decades can be extrapolated, this will certainly be available sometime in the next century. Though most of us will still be tapping out numbers in 2001, I've little real doubt that well before 2076 you will simply say to your comsole: "Get me Bill Smith." Or if you *do* say: "Get me 212-345-5512," it will answer, "Surely you mean 212-345-55*21*." And it will be quite right.

Now, a machine with this sort of capability—a robot secretary, in effect—could be quite expensive. *It doesn't matter.* We who are living in an economic singularity—if not a fiscal black hole—have forgotten what most of history must be like.

Contrary to the edicts of Madison Avenue, the time will come when it won't be necessary to trade in last year's model. Eventually, everything reaches its technological plateau, and thereafter the only changes are in matters of style. This is obvious when you look at such familiar domestic objects as chairs, beds, tables, knives, forks. Oh, you can make them of plastic or fiberglass or whatever, but the basic design rarely alters.

It took a few thousand years to reach these particular plateaus; things happen more quickly nowadays even for much more complex devices. The bicycle took about a century; radio receivers half that time. This is not to deny that marginal improvements will go on indefinitely, but after a while all further changes are icing on a perfectly palatable cake. You may be surprised to learn that there are electrical devices that have been giving satisfactory service for half a century or more. The other day someone found an Edison carbon filament lamp that has apparently never been switched off since it was installed. And until recently, there were sections of Atlantic cable that had been in service for a full century!

Now, it's hard to see how a properly designed and constructed solid-state device can ever wear out. It should have something like the working life of a diamond, which is adequate for most practical purposes. So when we reach this state of affairs, it would be worth investing more in a multipurpose home communications device than an automobile. It could be handed on from one generation to the next—as was once the case with a good watch.

It has been obvious for a very long time that such audio-visual devices could complete the revolution started by the telephone. We are already approaching the point when it will be feasible—not necessarily desirable—for those engaged in what are quaintly called "white collar" jobs to do perhaps 95 percent of their work without leaving home. Of course, few of today's families could survive this, but for the moment let's confine ourselves to electronic, not social, technology.

Many years ago I coined the slogan "Don't commute —communicate!" Apart from the savings in travel time (the *real* reason I became a writer is that I refuse to spend more than 30 seconds moving from home to office), there would be astronomical economies in power and raw materials. Compare the amount of hardware in communications systems, as opposed to railroads, highways and airlines. And the number of kilowatt hours you expend on the shortest journey would power several lifetimes of chatter between the remotest ends of the Earth.

Obviously, the home comsole would handle most of today's first-class mail; messages would be stored in its memory, waiting for you to press the playback key whenever you felt like it. Then you would type out the answer—or alternatively call up the other party for a face-to-face chat.

Fine, but at once we have a serious problem—the already annoying matter of time zones. They are going to become quite intolerable in the electronic global village—where we are all neighbors but a third of us are asleep at any given moment. The other day I was woken up at 4:00 A.M. by the London *Daily Express*, which had subtracted 5½ hours instead of adding them. I don't know what I said, but I doubt if my views on the Loch Ness Monster were printable.

The railroads and the telegraph made time zones inevitable in the nineteenth century; the global telecommunications network of the twenty-first may abolish them. It's been suggested, at least half seriously,[6] that we'll have to establish a Common Time over the whole planet—whatever inconvenience this may cause to those old-fashioned enough to gear themselves to the day-night cycle.

During the course of the day—whatever *that* may be—you will use the home comsole to call your friends and deal with business, exactly as you use the telephone now—with this difference. You'll be able to exchange any amount of tabular, visual or graphical information. Thus if you're an author, you'll be able to wave that horrid page one type in front of your delinquent editor on Easter Island, or wherever he lives. Instead of spending hours hunting for nonexistent parts numbers, engineers will be able to *show* their supplier the broken dohickey from the rotary discombobulator. And we'll all be able to see those old friends of a lifetime, whom we'll never again meet in the flesh.

Which raises an interesting problem. One of the great advantages of Mr. Bell's invention is that you can converse with people *without* their seeing you, or knowing where you are, or who is with you. A great many business deals would never be consummated, or even attempted, over a video circuit; but perhaps they are deals that shouldn't be, anyway. . . .

I am aware that previous attempts to supply vision—such as the Bell Picturephone—have hardly been a roaring success. But I feel sure that this is due to cost, the small size of the picture, and the limited service available. No one would have predicted much of a future for the very first "Televisors," with their flickering, postage-stamp–sized images. Such technical limitations have a habit of being rather rapidly overcome, and the *large screen, high definition* Picturephone-Plus is inevitable.

I could certainly do with such a device. For several years, Stanley Kubrick has been talking wistfully to me about another space project. But there's an insoluble problem—I won't leave Sri Lanka for more than a couple of weeks a year, and Stanley refuses to get into an airplane. We may both be too old, or too lazy, before the arrival of home comsoles makes another collaboration possible. So the present backwardness of electronics has spared the world another masterpiece.

Clearly, when we do have two-way vision, there will have to be some changes in protocol. You can't *always* pretend to your wife that the camera has broken down again. . . . Incidentally, some of the changes that would be produced in a society totally orientated to telecommunications have been well discussed by a promising local writer, in a novel called *The Naked Sun*. The author's full name escapes me at the moment, but I believe it begins with "Isaac."

The possibilities of the comsole as an entertainment and information device are virtually unlimited; some of them, of course, are just becoming available, as an adjunct to the various TV subscription services. At any moment one should be able to call up all the news headlines on the screen, and expand any of particular interest into a complete story at several levels of thoroughness—all the way, let us say, from the *Daily News* to *The New York Times* . . . I hate to think of the hours I have wasted listening to the radio news bulletins—for some item that never turned up. Nothing is more frustrating—as will be confirmed by any Englishman touring the United States during a Test Match, or any American in England during the World Series (how did it get that ridiculous name?). For the first time, it will be possible to have a news service with immediacy, selectivity *and* thoroughness.

The electronic newspaper, apart from all its other merits, will also have two gigantic ecological plusses. It will save whole forests for posterity; and it will halve the cost of garbage collection. This alone might be enough to justify it, and to pay for it.

Like many of my generation, I became a news addict during the Second World War. Even now, it takes a definite effort of will for me *not* to switch on the hourly news summaries, and with a truly global service one could spend every waking minute monitoring the amusing, crazy, interesting and tragic things that go on around this planet. I can foresee the rise of even more virulent forms of news addiction, resulting in the evolution of a class of people who can't bear to miss anything that's happening, anywhere, and spend their waking hours glued to the comsole. I've even coined a name for them—Infomaniacs.

Continuing in this vein, I used to think how nice it would be to have access, in one's own home, to all the books and printed matter, all the recordings and movies, all the visual arts of mankind. But would not many of us be completely overwhelmed by such an embarrassment of riches, and solve the impossible problem of selection by selecting nothing? Every day I sneak guiltily past my set of the Great Books of the Western World, most of which I've never even opened. . . . What would it *really* be like to have the Library of Congress—*all* the world's great libraries—at your fingertips? Assuming, of course, that your fingertips were sufficiently educated to handle the problem of indexing and retrieval. I speak with some feeling on this subject, because for a couple of years I had the job of classifying and indexing everything published in the physical sciences, in all languages. If you can't find what you're looking for in *Physics Abstracts* for 1949–51, you'll know who to blame.

With the latest techniques, it would be possible to put the whole of human knowledge into a shoebox. The problem, of course, is to get it out again; anything misfiled would be irretrievably lost. Another problem is to decide whether we mass-produce the shoeboxes, so that every family has one—or whether we have a central shoebox linked to the home with wide-band communications.

Probably we'll have both, and there are also some interesting compromises. Years ago I invented something that I christened, believe it or not, the *Micropaedia Britannica* (recently I was able to tell Mortimer Adler that I'd thought of the name first). But my *Micropaedia* would be a box about the size of an ordinary hard-cover book, with a display screen and alpha-numeric keyboard. It would contain, in text and pictures, *at least* as much material as a large encyclopaedia plus dictionary.

However, the main point of the electronic *Britannica* would not be its compactness—but the fact that, every few months, you could plug it in, dial a number, and have it updated overnight. . . . Think of the saving in wood pulp and transportation that this implies!

It is usually assumed that the comsole would have a flat TV-type screen, which would appear to be all that is necessary for most communications purposes. But the ultimate in face-to-face electronic confrontation would be when you could not tell, without touching, whether or not the other person was physically present; he or she would appear as a perfect 3D projection. This no longer appears fantastic, now that we have seen holographic displays that are quite indistinguishable from reality. So I am sure that this will be achieved someday; I am not sure how badly we need it.

What *could* be done, even with current techniques, is to provide 3D—or at least wide-screen Cinerama-type —pictures for a single person at a time. This would need merely a small viewing booth and some clever optics, and it could provide the basis for a valuable educational-entertainment tool, as Dennis Gabor has suggested.[7] But it could also give rise to a new industry— personalized television safaris. When you can have a high-quality cinema display in your own home, there will certainly be global audiences for specialized programs with instant feedback from viewer to cameraman. How nice to be able to make a trip up the Amazon, with a few dozen unknown friends scattered over the world, with perfect sound and vision, being able to ask your guide questions, suggest detours, request closeups of interesting plants or animals—in fact, sharing everything except the mosquitoes and the heat!

It has been suggested that this sort of technology might ultimately lead to a world in which no one ever bothered to leave home. The classic treatment of this theme is, of course, E. M. Forster's *The Machine Stops*, written more than 70 years ago as a counterblast to H. G. Wells.

Yet I don't regard this sort of pathological, sedentary society as very likely. "Telesafaris" might have just the opposite effect. The customers would, sooner or later, be inspired to visit the places that really appealed to them . . . mosquitoes notwithstanding. Improved communications will promote travel for *pleasure;* and the sooner we get rid of the other kind, the better.

So far, I have been talking about the communications devices in the home and the office. But in the last few decades we have seen the telephone begin to lose its metal umbilical cord, and this process will accelerate. The rise of walkie-talkies and Citizen's Band radio is a portent of the future.

The individual, wristwatch telephone through which you can contact anyone, anywhere, will be a mixed blessing which, nevertheless, very few will be able to reject. In fact, we may not have a choice; it is all too easy to imagine a society in which it is illegal to switch off your receiver, in case the Chairman of the People's Cooperative wants to summon you in a hurry. . . . But let's not ally ourselves with those reactionaries who look only on the *bad* side of every new development. Alexander Graham Bell cannot be blamed for Stalin, once aptly described as: "Genghis Khan with a telephone."

It would be an *underestimate* to say that the wristwatch telephone would save tens of thousands of lives a year. Everyone of us knows of tragedies—car accidents on lonely highways, lost campers, overturned boats, even old people at home—where some means of communication would have made all the difference between life and death. Even a simple emergency SOS system, whereby one pressed a button and sent out a HELP! signal, would be enough. This is a possibility of the immediate future; the only real problem—and, alas, a serious one—is that of false alarms.

At this point, before I lose all credibility with the hairy-knuckled engineers who have to produce the hardware, I'd better do a once-over-lightly of the electromagnetic spectrum. This is, I think, unique among our natural resources. We've been exploiting it for less than one lifetime and are now polluting much of it to the very maximum of our ability. But if we stopped using it tomorrow, it would be just as good as new, because the garbage is heading outward at the speed of light. . . . Too bad this isn't true of the rest of the environment.

Do we have enough available bandwith for a billion personal transceivers, even assuming that they aren't all working at once? As far as the home equipment is concerned, there is no problem, at least in communities of any size. The only uncertainty, and a pretty harrowing one to the people who have to make the decisions, is how quickly coaxial cables are going to be replaced by glass fibers, with their million-fold-greater communications capability. Incidentally, one of the less glamorous occupations of the future will be mining houses for the rare metal, copper, buried inside them by our rich ancestors. Fortunately, there is no danger that we shall ever run out of silica. . . .

But I would also suggest that optical systems, in the infrared and ultraviolet, have a great future not only for fixed, but even for *mobile,* personal communications. They may take over some of the functions of present-day transistor radios and walkie-talkies—leaving the radio bands free for services which can be provided in no other way. The fact that opticals have only very limited range, owing to atmospheric absorption, can be turned to major advantage. You can use the same frequencies—and *what* a band of frequencies! —millions of times over—as long as you keep your service areas 10 or 20 kilometers apart.

It may be objected that light waves won't go round corners, or through walls. Elementary, my dear Watson. We simply have lots of dirt-cheap—because they are made from dirt!—optical wave guides and light pipes deliberately leaking radiation all over the place. Some would be passive, some active. Some would have very low-powered optical-to-radio transducers in both directions, to save knocking holes in walls, and to get to awkward places. In densely populated communities one would always be in direct or reflected sight of some optical transmitter or repeater. But we must be careful how we use the ultraviolet. People who talked too much might get sunburned. . . .

When you are cycling across Africa, or drifting on a balsa-wood raft across the Pacific, you will of course still have to use the radio frequencies—say the one to ten thousand megahertz bands, which can accommodate at least 10 million voice circuits. This number can be multiplied many times by skillful use of satellite technology. I can envisage an earth-embracing halo of low-altitude, low-powered radio satellites, switching frequencies continually so that they provide the desired coverage in given geographical regions. And NASA has recently published a most exciting report on the use of the very large (kilometer-square!) antennas we will soon be able to construct in space.[8] These would permit the simultaneous use of myriads of very narrow beams which could be focused on individual subscribers, carrying receivers which could be mass-produced for about 10 dollars. I rather suspect that our long-awaited personal transceiver will be an adaptive, radio-optical hybrid, actively hunting the electromagnetic spectrum in search of incoming signals addressed to it.

Now, the invariably forgotten accessory of the wrist-watch telephone *directory*. Considering the bulk of that volume for even a modest-sized city, this means that our personal transceivers will require some sophisticated information-retrieval circuits, and a memory to hold the few hundred most-used numbers. So we may be forced, rather quickly, to go the whole way and combine in a single highly portable unit not only communications equipment, plus something like today's pocket calculators, plus data banks, plus information processing circuits. It would be a constant companion, serving much the same purpose as a human secretary. In a recent novel I called it a "Minisec." [9] In fact, as electronic intelligence develops, it would provide more and more services, finally developing a personality of its own, to a degree which may be unimaginable today.

Except, of course, by science fiction writers. In his brilliant novel *The Futurological Congress*, Stanislaw Lem gives a nightmare cameo which I can't get out of my mind. He describes a group of women sitting in complete silence—while their handbag computers gossip happily to one another. . . .

One of the functions of science fiction is to serve as an early warning system. In fact, the very act of description may prevent some futures, by a kind of exclusion principle. Far from predicting the future, science fiction often *exorcises* it. At the very least, it makes us ask ourselves: "What kind of future do we really want?" No other type of literature poses such fundamental questions, at any rate explicitly.

The marvelous toys that we have been discussing will simply remain toys, unless we use them constructively and creatively. Now, toys are all right in the proper place; in fact they are an essential part of any childhood. But they should not become mere distractions—or ways of drugging the mind to avoid reality.

We have all seen unbuttoned beer-bellies slumped in front of the TV set, and transistorized morons twitching down the street, puppets controlled by invisible disk jockeys. These are not the highest representatives of our culture; but, tragically, they may be typical of the near future. As we evolve a society orientated toward information, and move away from one based primarily on manufacture and transportation, there will be millions who cannot adapt to the change. We may have no alternative but to use the lower electronic arts to keep them in a state of drugged placidity.

For in the world of the future, the sort of mindless labor that has occupied 99 percent of mankind, for much more than 99 percent of its existence, will of course be largely taken over by machines. Yet most people are bored to death without work—even work that they don't like. In a workless world, therefore, only the highly educated will be able to flourish, or perhaps even to survive. The rest are likely to destroy themselves and their environment out of sheer frustration. This is no vision of the distant future; it is already happening, most of all in the decaying cities.

So perhaps we should not despise TV soap operas if, during the turbulent transition period between our culture and real civilization, they serve as yet another opium for the masses. *This* drug, at any rate, is cheap and harmless, serving to kill Time—for those many people who like it better dead.

When we look at the manifold problems of our age, it is clear that the most fundamental one—from which almost all others stem—is that of ignorance. And ignorance can be banished only by communication, in the widest meaning of the word.

The best educational arrangement, someone once remarked, consists of a log with a teacher at one end and a pupil at the other. Unfortunately there are no longer enough teachers, and probably not enough logs, to go around.

Now, one thing that electronics can do rather well is

to multiply teachers. As you doubtless know, at this very moment a most ambitious and exciting social experiment is taking place in India, where NASA's ATS-6 satellite is broadcasting educational programs to several thousand villages. ATS-6 is the only communications satellite in existence powerful enough to transmit signals that can be picked up on an ordinary TV set, augmented by a simple parabolic dish, like a large umbrella made of wire mesh.

Thanks to the extraordinary generosity of the Indian Space Research Organization, which flew in six engineers and half a ton of equipment, I have a five-meter satellite antenna on the roof of my Colombo house, now renamed Jodrell Bank East. Since the experiment started on 1 August 1975, I have thus been in the curious position of having the only TV set in the country. It's been fascinating to watch the programs; even though I don't understand Hindi, the messages of family planning, hygiene, agricultural techniques and national unity come across loud and clear.

Though it is impossible to put a value on such things, I believe that the cost of this experiment will be trivial compared with the benefits. And the ground segment is remarkably cheap, in terms of its coverage. Would you believe four thousand people round one TV set? Or a three-meter-diameter village antenna—made of *dried mud?*

Of course, there are some critics—as reported recently by Dr. Yash Pal, the able and energetic Director of the Indian Space Application Centre: [10]

"In the drawing rooms of large cities you meet many people who are concerned about the damage one is going to cause to the integrity of rural India by exposing her to the world outside. After they have lectured you about the dangers of corrupting this innocent, beautiful mass of humanity, they usually turn round and ask:

'Well, now that we have a satellite, when are we going to see some American programs?' Of course, they themselves are immune to cultural domination or foreign influence."

I'm afraid that cocktail party intellectuals are the same everywhere. Because *we* frequently suffer from the modern scourge of information pollution, we find it hard to imagine its even deadlier opposite—information starvation. For any outsider, however well-meaning, to tell an Indian villager that he would be better off without access to the world's news, entertainment *and knowledge* is an obscene impertinence, like the spectacle of a fat man preaching the virtues of fasting to the hungry.

Unfortunately, on 31 July 1976 the one-year experiment will end; ATS-6 will crawl back along the equator and return to the United States. Originally, it was hoped to launch *two* satellites; last summer I saw the three-quarters-completed ATS-7, sitting mothballed at the Fairchild plant. No one could raise the 10 million necessary to finish it, or hijack one of the Air Force's numerous Titan 3C's to get it into orbit.

And so in a few month's time, millions of people who have had a window opened on marvelous new worlds of culture and education will have it slammed shut in their faces. There will be some heartrending scenes in the villages, when the cry goes up, however unfairly, "The Americans have stolen our satellite!" Useless to explain, as the frustrated viewers start to refill their six-to-nine time slot with baby making, that it was only through the initiative and generosity of the United States that the satellite was loaned in the first place. . . . The Ugly American will have struck again.

Yet I hope that this noble experiment is just the curtain raiser to a truly global educational satellite system.

Its cost would be one or two dollars per student, per *year*. There could be few better investments in the future health, happiness and peace of mankind.

I don't wish to get too much involved in the potential —still less the politics—of communications satellites, because they can take care of themselves, and are now multiplying rapidly. The world investment in satellites and ground stations now exceeds a billion dollars and is increasing almost explosively. After years of delay and dithering, the United States is at last establishing *domestic* satellite systems; the U.S.S.R. has had one for almost a decade. At first, the Soviet network employed *non*synchronous satellites, moving in an elongated orbit that took them high over Russia for a few hours of every day. However, they have now seen the overwhelming advantages of stationary orbits, and several of their comsats are currently fixed above the Indian Ocean. Some are designed for TV relaying to remote parts of the Soviet Union, and I've gently hinted to my friends in Moscow that perhaps *they* could fill the breach when ATS-6 goes home. . . .

We are now in the early stages of a battle for the mind—or at least the eyes and ears—of the human race, a battle which will be fought 36,000 kilometers above the equator. The preliminary skirmishes have already taken place at the United Nations, where there have been determined attempts by some countries to limit the use of satellites which can beam programs from space directly into the home, thus bypassing the national networks. Guess who is scared. . . .

As a matter of fact, I tried to frighten the United States with satellites myself, back in 1960, when I published a story in *Playboy* [11] about a Chinese plot to brainwash innocent Americans with pornographic TV programs. Perhaps "frighten" is not the correct verb, and in these permissive days such an idea sounds positively old-fashioned. But in 1960 the first regular com-sat service was still five years in the future, and this seemed a good gambit for attracting attention to its possibilities.

Fortunately, in this area there is an excellent record of international cooperation. Even countries who hate each other's guts work together through the International Telecommunications Union, which sets limits to powers and assigns frequencies. Eventually, some kind of consensus will emerge, which will avoid the worst abuses.

A major step toward this was taken on August 20, 1971, when the agreement setting up INTELSAT (the International Telecommunications Satellite Organization) was signed at the State Department. I would like to quote from the address I gave on that occasion:

I submit that the eventual impact of the communications satellite upon the whole human race will be at least as great as that of the telephone upon the so-called developed societies.

In fact, as far as real communications are concerned, there are as yet no developed societies; we are all still in the semaphore and smoke-signal stage. And we are now about to witness an interesting situation in which many countries, particularly in Asia and Africa, are going to leapfrog a whole era of communications technology and go straight into the space age. They will never know the vast networks of cables and microwave links that this country has built at such enormous cost both in money and in natural resources. The satellites can do far more and at far less expense to the environment. . . .

. . . I believe that the communications satellites can unite mankind. Let me remind you, that, whatever the history books say, this great country was created a little more than a hundred years ago by two inventions. Without them, the United States was impossible; with them, it was inevitable. Those inventions were, of course, the railroad and the electric telegraph.

Today we are seeing on a global scale an almost exact parallel to that situation. What the railroads and the telegraph did here a century ago, the jets and the communications satellites are doing now to all the world. . . .

And the final result—whatever name we actually give to it—will be the United States of Earth.

I would like to end with some thoughts on the wider future of communications—communications beyond the earth. And here we face an extraordinary paradox, which in the centuries to come may have profound political and cultural implications.

For the whole of human history, up to that moment 100 years ago when the telephone was invented, it was impossible for two persons more than a few meters apart to interact in real time. The abolition of that apparently fundamental barrier was one of technology's supreme triumphs; today we take it for granted that men can converse with each other, and even see each other, wherever they may be. Generations will live and die, always with this godlike power at their fingertips.

Yet this superb achievement will be ephemeral; before the next 100 years have passed, our hard-won victory over space will have been lost, never to be regained.

On the Apollo voyages, for the first time, men traveled more than a light-second away from Earth. The resulting 2½-second round-trip delay was surprisingly unobtrusive, but only because of the dramatic nature of the messages—and the discipline of the speakers. I doubt if the average person will have the self-control to talk comfortably with anyone on the Moon.

And beyond the Moon, of course, it will be impossible. We will never be able to converse with friends on Mars, even though we can easily exchange any amount of information with them. It will take at least three minutes to get there, and another three minutes to receive a reply.[12]

Anyone who considers that this is never likely to be of much practical importance is taking a very short-sighted view. It has now been demonstrated, beyond reasonable doubt, that in the course of the next century, we could occupy the entire Solar System. The resources in energy and material are there; the unknowns are the motivation—and our probability of survival, which may indeed depend upon the rate with which we get our eggs out of this one fragile planetary basket.

We would not be talking about the future unless we were optimists. And in that case we must *assume* that eventually very large populations will be living far from Earth—light-minutes and light-hours away, even if we only colonize the inner Solar System. However, Freeman Dyson has argued with great eloquence [13] that planets aren't important, and the real action will be in the cloud of comets out beyond Pluto, a light *day* or more from earth.

And looking further afield, it is now widely realized that there are no *fundamental* scientific obstacles even to interstellar travel.[14] Though Dr. Purcell once rashly remarked that star-ships should stay on the cereal boxes, where they belonged—that's exactly where moonships were, only 30 years ago . . .

So the finite velocity of light will, inevitably, divide the human race once more into scattered communities, sundered by barriers of space and time. We will be as one with our remote ancestors, who lived in a world of immense and often insuperable distances, for we are moving out into a universe vaster than all their dreams.

But it is, surely, not an empty universe. No discussion of communications and the future would be complete without reference to the most exciting possibility of all—communications with extraterrestrial intelligence. The Galaxy must be an absolute Babel of conversation, and it is surely only a matter of time before we can hear the neighbors. They already know about us, for our sphere of detectable radio signals is now scores of light-years across. Perhaps even more to the point—and more likely to bring the precinct cops hurrying here as fast as their paddy wagon can travel— is the fact that several microsecond-thick shells of X-ray pulses are already more than 10 light-years out from earth, announcing to the universe that, somewhere, juvenile delinquents are detonating atom bombs.

Plausible arguments suggest that our best bet for interstellar eavesdropping would be in the 1,000 megahertz, or 30 centimeter, region. The NASA/Stanford/ Ames *Project Cyclops* report, which proposed an array of several hundred large radio telescopes for such a search, recommended a specific band about 200 megahertz wide—that lying between the hydrogen line (1420 MHz) and the lowest OH line (1662 MHz). Dr. Bernard Oliver, who directed the Cyclops study, has waxed poetic about the appropriateness of *our* type of life seeking its kind in the band lying between the disassociation products of water—the "water hole." [15]

Unfortunately, we may be about to pollute the water hole so badly that it will be useless to radio astronomers. The proposed MARESAT and NAVSTAR satellites will be dunked right in the middle of it, radiating so powerfully that they would completely saturate any Cyclops-type array. Barney Oliver tells me: "Since the Cyclops study, additional reasons have become apparent for expecting the water hole to be our contact with the mainstream of life in the Galaxy. The thought that we, through our ignorance, may blind ourselves to such contact and condemn the human race to isolation appalls us."

I hope that the next World Administrative Radio Conference, when it meets in 1979, will take a stand on this matter. The conflict of interest between the radio astronomers and the communications engineers will get more and more insoluble, until, as I suggested many years ago,[16] we move the astronomers to the quietest place in the Solar System—the center of the lunar Farside, where they will be shielded from the radio racket of Earth by 3,500 kilometers of solid rock. But *that* answer will hardly be available before the next century.

Whatever the difficulties and problems, the search for extraterrestrial signals will continue. Some scientists fear that it will not succeed; others fear that it *will*. It may already have succeeded, but we don't yet know it. Even if the pulsars *are* neutron stars—so what? They may still be artificial beacons, all broadcasting essentially the same message: "Last stop for gas this side of Andromeda."

More seriously, if the decades and the centuries pass, with no indication that there is intelligent life elsewhere in the universe, the long-term effects on human philosophy will be profound—and may be disastrous. Better to have neighbors we don't like than to be utterly alone. For that cosmic loneliness could point to a very depressing conclusion—that intelligence marks an evolutionary dead-end. When we consider how well —and how *long*—the sharks and the cockroaches have managed without it, and how badly we are managing *with* it, one cannot help wondering if intelligence is an aberration like the armor of the dinosaurs, dooming its possessors to extinction.

No, I don't *really* believe this. Even if the computers we carry on our shoulders are evolutionary accidents, they can now generate their own programs—and set their own goals.

For we can now say, in the widest possible meaning of the phrase, that the purpose of human life is information processing. I have already mentioned the strange fact that men can survive longer without water than without information. . . .

And therefore the real value of all the devices we have been discussing is that they have the potential for immensely enriching and enlarging life, by giving us more information to process—up to the maximum number of bits per second that the human brain can absorb.

I am happy, therefore, to have solved one of the great problems the philosophers and theologians have been haggling over for several thousand years. You may, perhaps, feel that this is rather a dusty answer, and that not even the most inspired preacher could ever found a religion upon the slogan: "The purpose of life is information processing." Indeed, you may even retort: "Well, what is the purpose of information processing?"

I'm glad you asked me that . . .

References

1. *Glide Path* (New York: Harcourt Brace Jovanovich, 1963). See also "You're on the glide path—I think" (*The Aeroplane*, Sept. 23, 1949. Reprinted in *IEEE Transactions on Aerospace and Navigational Electronics*, Vol. ANE-10, No. 2, June 1963.) There is an echo of G.C.A. in my 1945 communications satellite paper. The S-band Search radar had a cosecant-squared polar diagram to increase field strength at slant range. I realized that comsats would face a similar problem and therefore suggested the use of a "non-uniform radiator." As the war was still on, and my MS had to be checked by RAF Security, this was sailing as close to the wind as I dared.

2. *The New York Times* Arts and Leisure (music section), May 11, 1975.

3. I am much indebted to Anthony Wedgwood-Benn for this quotation, which he discovered when he was H.M. Postmaster General. However, to do proper justice to Sir William Preece, it should also be pointed out that he was largely responsible for setting up the British telephone system—and was also a quarter of a century later, a pioneer of "wireless" and one of Marconi's chief supporters!

4. *Imperial Earth* (New York: Harcourt Brace Jovanovich, 1975).

5. No joke, this. My friends at Time-Life once told me that when they edited Churchill's war memoirs, they found countless statistical errors. The trouble was traced to the great man's habit of dentureless dictation.

6. By Dr. Solomon Golumb.

7. *The Mature Society* (New York: Praeger Publishers, 1972).

8. Aerospace Corporation Report, *Potential Space System Contributions in the Next Twenty Five Years*, 1975. For summary, see Vol. II of the House of Representatives Subcommittee of Space Science and Applications, Future Space Programs, 1975.

9. *Imperial Earth* (*op. cit.*).

10. "Some lessons during the setting up of SITE." (Talk at UN/UNESCO Regional Seminar on Satellite Broadcasting Systems for Education and Development, Mexico City, Sept. 2–11, 1975.)

11. "I Remember Babylon." (*Playboy*, May 1960. Reprinted in *Tales of Ten Worlds* [New York: Harcourt Brace Jovanovich, 1962].)

12. See "Don't write; telegraph" by J. J. Coupling (*Astounding Science Fiction*, March 1952). Mr. Coupling was a promising science fiction writer whose output was sadly limited by the activities of his *alter ego*, Dr. John Pierce, Director of Communications Research at Bell Labs.

13. "The World, the Flesh and the Devil," Third J. D. Bernal Lecture, Birkbeck College, London, 1972. Now available as Appendix D to *Communications with Extraterrestrial Intelligence*, ed. by Carl Sagan (Cambridge: M.I.T. Press, 1973).

14. See, for example, the "Interstellar Studies" issues of the *Journal of the British Interplanetary Society*. Just 40 years ago, amid general incredulity, *J.B.I.S.* started to publish studies of vehicles which could carry men to the Moon. This is where we came in. . . .

15. *Project Cyclops:* A Design Study of a System for Detecting Extraterrestrial Intelligent Life (NASA/Ames CR 114445).

16. "The Uses of the Moon," *Harper's*, December 1961. Reprinted in *Voices from the Sky* (New York: Harper & Row, 1965).

AFTERWORD
John D. deButts

One hundred years ago, Alexander Graham Bell and his associate Thomas Watson were working in Bell's rooms at 5 Exeter Place—about 10 blocks from the locale of our convocation—in yet another effort, among so many others that had proven vain, to make the telephone work. Bell was in his workshop manipulating his liquid transmitter, which was connected by wire to a bedroom where Watson stood by a tuned reed receiver, hoping to catch whatever faint emission it might produce.

What gave rise to the first transmission of intelligible speech was an accident. Mr. Bell inadvertently spilled some acid on himself and—without any notion that they would live in history—spoke these historic words: "Mr. Watson, come here, I want you!"

Watson had no transmitter, and Bell no receiver. Nonetheless, that first emergency call got a quick response. Watson rushed down the hall, burst into Bell's workshop and told him: "Mr. Bell, I heard every word you said distinctly." "Distinctly" isn't quite the way I would describe how Mr. Watson heard Mr. Bell one hundred years ago. At any rate, he heard, he understood, and he came running.

Mr. Bell's life and Mr. Watson's life took separate paths. Almost 40 years later, however—in 1915—they were once again on opposite ends of a telephone line. And once again Mr. Bell uttered his historic words:

"Mr. Watson, come here, I want you." This time Mr. Bell was in New York, the central figure in the ceremony inaugurating the Bell System's first coast-to-coast telephone service. Mr. Watson was almost 3,000 miles away—in San Francisco. Once again he told Mr. Bell that he heard every word distinctly. "But if you want to see me," he added, "it will take me almost a week to get there!"

The first transcontinental call traveled by open wire from coast to coast. Since then, we have developed transmission systems of greater and greater capacity—and economy. In 1915 a coast-to-coast call cost $20.70. Now you can call for as little as 21 cents—the rate for a one-minute dial-it-yourself call at night or on weekends.

In the 1920's and early 1930's we developed the coaxial cable, and began putting it into service just before World War II. Today's coaxial cable systems can handle as many as 108,000 simultaneous conversations, compared to 480 in the first coaxial cables.

After World War II came microwave radio relay. Today systems like this account for about 70 per cent of the circuit mileage in the interstate telephone network.

In 1956 we laid down our first transatlantic cable, and the cable ship "Long Lines" is now steaming toward mid-ocean laying down Transatlantic Cable Number Six.

In 1962 the Bell System inaugurated the age of satellite communications with Telstar. In May 1976 the Bell System, together with General Telephone and Electronics, will begin using satellites for domestic communications, thereby endowing the nationwide switched network with still another dimension of flexibility.

Today we are readying transmission systems of still

greater capacity. For example, the millimeter wave-guide, when we need it, will permit us to carry more than a quarter of a million simultaneous conversations in a hollow "pipe" no thicker than a man's fist. I say "when we need it" because not until the 1980's do we anticipate the traffic volumes that will warrant its installation on a commercial basis. When the need comes, however, we'll be ready.

Until recently most of us thought of lightwave communications as a 21st-century prospect. Now it seems likely to be an operating reality in our business in the relatively near future. Already lightwave communications has passed most of the hard tests of practical application, and I anticipate that by the early 1980's cables of glass fibers will be carrying thousands of simultaneous messages between major switching centers in our big cities.

To communicate is the beginning . . . Of what, no one of us can be sure. But this much we can say. No technology serves ends more profoundly human than does the technology of communications. And beyond a doubt it was to human ends that Alexander Graham Bell dedicated his inventive genius.

What seems to me most remarkable about the telephone's invention is that its inventor—before he was an inventor—was a teacher. Bell—for all his lifelong interest in things mechanical and electrical—was not a scientist. He was not a technologist in the sense that we understand the term today. He came to the telephone out of his efforts to teach deaf children to speak. It was not the tinkerer's urge that spurred him. It was a basic human need.

One recalls the story of Bell's complaint to the celebrated physicist Joseph Henry, then 77 years old, that he lacked the knowledge of electrical principles to bring his idea to practical fruition. The old man's response was brief and blunt. "Get it," he said.

And get it Bell did—at least a sufficiency to translate what had hitherto been but a vaguely grasped notion into one of the landmark inventions in all of human history. Having done that, he went on to other things—to airplanes and hydrofoils and kites, and always and forever the teaching of the deaf to speak. It remained for others to develop the potentialities latent in U.S. Patent Number 174,465.

Before he died, however, he left for his successors a shaping vision—a vision as intensely human as the vision that drove him to the telephone's invention in the first place. From London in 1878 he wrote to prospective investors in the telephone as follows: "I believe, in the future, wires will unite the head offices of the Telephone Companies in different cities, and a man in one part of the Country may communicate by word of mouth with another in a distant place. . . . I would impress upon you all the advisability of keeping this end in view, that all present arrangements of the Telephone may eventually be realized in this grand system."

"To another in a distant place"—as many times as I have read those words there remains a high excitement in them that speaks of far horizons and still farther horizons beyond those to a limit beyond man's knowing. In those words is summed up all that's been done in telephony since Bell's invention and all that we, the heirs to his discovery, yet seek to do.

It has been said—and on this centennial it is being said again—that Alexander Graham Bell's vision has been fulfilled. To that I say no.

Looking ahead, I am convinced that we are a long way from achieving the universality of communications that our technology can accomplish. Indeed, I believe that technology now being readied for application will create in telephony's second century an era of ever more abundant communications, and thereby vastly extend man's reach, expand his ability to manage complex undertakings, and—not least of all—enlarge his freedoms.

Three factors, I believe, will characterize the century now beginning: (1) the sheer abundance of communications; (2) its diversity. Increasingly the nationwide switched network will enable us to communicate with others in distant places not only by word of mouth but in pictures or in the language of computers or in virtually whatever format information can be arranged; and (3) telephony's second century will differ from its first in its global reach. Already the U.S. telephone system is linked with the telephone systems of 212 of the world's nations. As the years unfold, I expect to see—in some cases swiftly, in others haltingly—the whole world's telephone system moving toward the level of advanced development that we in the United States enjoy today. How soon this will happen depends more on politics than it does on technology. But when it comes, as it surely will, I cannot help but feel that it will be a benign development. Communications, I believe, is the beginning of community.

Thus far we have been celebrating technology, its triumphs and its prospects. But it was not technology that shaped our industry's goals. Rather it was the other way around. The industry's goals shaped its technology. Accordingly, it seems to me that, along with the inventions and inventors that we commemorate, we should commemorate as well the contributions of those leaders of our business, not all of them technologists, who down through the years defined our industry's aims, established its operating principles, set its standards, and framed its organizational structure. Among those leaders, one stands pre-eminent in our memory. That one man can make a difference to the course of history seems to me clear when you consider how different a course our industry might have followed had Theodore Newton Vail not so compellingly established universal service as its goal, and the public interest as the prime standard of its performance. Vail, of course, was not alone in shaping the course of our industry's history, and there are a great many other leaders whose names I could recite who between his time and ours showed not only the wisdom to perceive but—in the face of challenge—the courage to speak out for what they believed was right. One can only hope that in the perspective of history those of us—not only in the industry, but in government as well—who have some share in resolving the current contentions besetting our industry will be perceived as having as large a view as theirs.

Finally, it seems to me that this celebration would be incomplete if we neglected to salute the generations of telephone men and women whose names are not in the history books and whose only monument is what they built. In our time one does not hear much of the spirit of service. That does not mean that it does not live. It lives because in ways I cannot define it was passed from generation to generation of telephone people, not only by precept but by example: the example of the craftsman who, aspiring perhaps to be nothing more, contented himself with being nothing less, a craftsman; the example of the operator who, in full knowledge of the millions of calls our industry handles every day, recognizes that each one, to someone, is more important than all the others. What we call the spirit of service was shaped on our industry's frontiers, its westward march across the searing deserts and the frigid passes traversed by our first transcontinental line. It was shaped in the blizzard of 1888 and in countless hurricanes and earthquakes and fires in which telephone people have braved discomfort and danger to keep the lines of communications open. Celebrating telephony's triumphs, let it not be said we failed to remember that there are men and women who gave their lives for it.

I don't know where the spirit of service originated. I know only that when I first came to the business almost 40 years ago, it was there and my first lessons were in what it meant. We have, I submit, no greater responsibility in our second century than to foster that heritage and, if we can, enhance it.

That much we owe to Alexander Graham Bell and the other towering figures in telephone history, who gave us not only our technology but the principles on which this industry was founded. We owe it to the hundreds of thousands of men and women who built the telephone system and brought it to its present standard of excellence. And we owe it to America and to our fellow Americans as together we enter the third century of this great nation with renewed hope and confidence.

We shall not fail.